TIME
THE REFRESHING RIVER

TIME
THE REFRESHING RIVER

TIME
THE REFRESHING RIVER

by
JOSEPH NEEDHAM
FRS, FBA

SPOKESMAN

First published in 1943 by George Allen and Unwin Ltd.
This edition published in 1986 by Spokesman,
Bertrand Russell House, Gamble Street,
Nottingham NG7 4ET.
Telephone 0602 708318

Distributed in the United States by:
Humanities Press International inc.
Atlantic Highlands, New Jersey 07716

British Library Cataloguing in Publication Data

Needham, Joseph
  Time: the refreshing river
  I. Title
  082   AC8

ISBN 0-85124-429-7
ISBN 0-85124-439-4 Pbk

Printed by the Russell Press Ltd., Gamble Street, Nottingham.
Telephone 0602 784505

CONTENTS

	PAGE
Preface to New Edition	i
Metamorphoses of Scepticism (1941)	7
The Naturalness of the Spiritual World: A Reappraisement of Henry Drummond (1939)	28
Science, Religion and Socialism (1935)	42
Laud, the Levellers, and the Virtuosi (1935)	75
Pure Science and the Idea of the Holy (1941)	92
Thoughts of a Young Scientist on the Testament of an Elder One (John Scott Haldane) (1936)	121
Limiting Factors in the History of Science, observed in the History of Embryology (Carmalt Lecture at Yale University, 1935)	141
The Biological Basis of Sociology (1936)	160
A Biologist's View of Whitehead's Philosophy (1941)	178
Evolution and Thermodynamics (1941)	207
Integrative Levels; A Revaluation of the Idea of Progress (Herbert Spencer Lecture at Oxford University, 1937)	233
Index	273

To
CONRAD NOEL
Priest of Thaxted
and
Prophet of Christ's Kingdom
on Earth

"Urbs Sion unica, mansio mystica, condita caelo,
Nunc tibi gaudio, nunc tibi lugeo, tristor, anhelo;
Te, quia corpore non queo, pectore saepe penetro,
Sed caro terrea, terraque carnea, mox cado retro.
Nemo retexere, nemoque promere sustinet ore
Quo tua moenia, quo capitolia plena nitore.
Id queo dicere, quo modo tangere pollice coelum
Ut mare currere, sicut in aere, figere telum.
Opprimit omne cor ille tuus decor, O Sion, O Pax,
Urbs sine tempore, nulla potest fore, laus tibi, mendax;
O nova mansio, te pia concio, gens pia munit,
Provehit, excitat, auget, identitat, efficat, unit."
from the Rhythm of Bernard of Cluny.
"Hora novissima," 12th century.

"Somewhere beyond the railheads
Of Reason, south or north,
Lies a Magnetic Mountain,
Riveting sky to earth."
C. Day Lewis

Preface

Needless to say, it is a very great pleasure for me to see "Time the Refreshing River" in print again after so many years. It was first published in a dark time, in 1943, and I finished correcting the proofs in China, where I had gone the year before to become Scientific Counsellor at the British Embassy in Chungking. Thus more than forty years have passed since the first appearance of this collection of essays. As this is a reprint, and not a new edition, no one will be looking for references in it later than 1943.

Many people all over the world have told me how much it influenced their thinking. I don't believe the essays in it are very much out-dated, partly because there are so many quotations from the 17th century and earlier times still. It contains what might be called a *philosophia perennis*.

This consisted, one might say, of two main themes. On the one hand, I had learnt from Rudolf Otto (the German theologian) and R.G. Collingwood (the Oxford philosopher) that there are five forms of human experience — science, religion, history, philosophy and artistic (or aesthetic) experience. Secondly, there was the conviction that evolution had been a real phenomenon which every religion, including Christianity, must come to terms with. On this view, cosmic evolution, inorganic evolution, organic evolution and social evolution were all one long, continuous process; and therefore the Kingdom of God on Earth was not some wild impossible dream, but something that had the whole authority of the evolutionary process behind it. The name of Teilhard de Chardin will not be found in the index of this book, but that was because I only came to know him in Paris at the end of the war, and even then I did not quite understand his writings until after his death. Still, I recognised a kindred spirit, if it would be allowable to speak in such a way about so charismatic a personality. Another great mentor will, however, be found there, namely Conrad Noel, whose teaching on the two different conceptions of *Regnum Dei* will certainly be found in this book. I first went over to Thaxted and listened to him in 1927 or so, and I am still going, so that is some sixty years of Christian socialism.

In my youth, I used to think that it didn't matter at all if the five forms of experience contradicted each other flat. I was thus some kind of existentialist (without knowing it), because I believed they could only be reconciled in a life lived. But now I have come to think that we need a much more unified world-view.

Partly I have been brought to feel this by the increase all over the world of theological "fundamentalism", which leads so easily to fanaticism. The idea of a literal inspiration of any people's holy scriptures is one which can do infinite harm. I fear it deeply, whether in Islam, where it leads to the re-introduction of barbarous punishments characteristic of the 7th century, or in Christendom where it induces whole populations to anticipate the Third World War as Armageddon, a conflict from which the elect will be miraculously saved. Even in normally peace-loving countries, fundamentalism and fanaticism can hardly be absent from the strife between the Buddhists and the Hindus. Fanaticism leads people to behave without any regard for their own safety, like the car-bombers of the Lebanon, or the Japanese *kamikaze* pilots of the last war; or indeed for the safety of others, however innocent. And as if air travel were not dangerous enough itself, the element of "hijacking" has to be added to it, again with great suffering to innocent passengers — as we have seen abundantly in recent cases.

In the past, there have been many dark ages, and we certainly seem to be living in one today. I am constantly appalled by the rôle of violence in the world at the present time. If the superpowers are for the moment refraining from using atomic weapons against each other, the world is full of undeclared wars, with even a few declared ones. Every American film that we see on British television exalts the use of the revolver, the modern "hand-gun"; and where that is not available, mobs go in for the kind of football hooliganism which leads not merely to injuries but many deaths. Young muggers think nothing of breaking into the homes of old people, tying them up and beating them, all for the sake of a few pounds of money. And then there is institutional violence — I say nothing of the conflicts of the police with the striking miners in England recently; but only a few days ago, we saw pictures on television of policemen breaking with their batons the windows of buses containing the hippies who wanted to have some kind of concert near Stonehenge. Everywhere in the world, armed conflict seems to be prevalent — one thinks especially of Nicaragua, El Salvador and Guatemala, but there is also the strife between Tamils and Singhalese in Ceylon, or the fighting between the authoritarian forces and the guerrillas in Peru, or the Afghan conflict and fighting still going on in Timor and New Caledonia. There seems to be nothing humane about civilisation today, no sense of restraint. But then the long-term view breaks in. There have been many very bad times in human history before; the Wars of Religion were appalling, and even in the eighteenth century there were mobs, footpads and highwaymen. But nevertheless social evolution has survived them all, and gone on from strength to strength. We can be quite confident that the same will happen again.

One of the most important events in all history was the invention of gunpowder, first made in 9th-century China and transmitted to Europe in the 13th century. The Chinese used it for war from the beginning of the 10th century onwards. But on the way they made some amazing discoveries, particularly the invention of the rocket, the only vehicle which can travel more easily in outer space than in the atmosphere.Moreover towards the end of the 17th century, before the steam-engine came into its own, people tried very hard to make gunpowder engines work. They failed, but it put them in mind of plain water and steam, so that Thomas Newcomen was able in 1712 to produce the first of all steam-engines. Thus this explosive played a fundamental part in the development of all heat engines.

Its social effects have often been meditated. A great Victorian writer, H.T. Buckle, saw its chief result in 1857 as the professionalisation of warfare. Gunpowder technology was complicated and difficult to handle. Therefore there arose inevitably a separate military profession, and ultimately standing armies; no longer was every man potentially a soldier. Hence there occurred a reduction in the proportion of the population entirely devoted to war, with the result that more people were shunted into peaceful arts, techniques and employment. Hence also what he called "a diminution of the warlike spirit by decreasing the number of persons for whom the practice of war was habitual." Gunpowder technology was also expensive, more so than any individuals could afford; so only wealthy republics, or kings backed by merchants and endowed with rich estates, could manufacture, own and operate musketry and artillery. Hence the rise of what Buckle called "the middle intellectual class" so that "the European mind, instead of being as heretofore solely occupied with either war or theology, now struck out into a middle path, and created those great branches of knowledge to which modern civilisation owes its origin."

As a description of one aspect of the rise of the bourgeoisie, this was all well said. But the only mistake made by Victorian optimism was the belief that the situation would last. Already towards the end of the 17th century, Robert Boyle, in describing the mechanism of the musket, had said that a single touch on the trigger could mean life or death. And in the course of time that touch would be open to everyone. It might have been wiser to foresee that science and technology would, as time went on, by the very impetus of the Industrial Revolution itself, which Buckle so much admired, immensely improve and enormously cheapen the production of these lethal weapons; not only on the mechanical side, but also on the chemical, producing a vast variety of explosives which could come within the reach of almost every man. Today, mass-produced firearms have become available to him, outstanding for their cheapness, range and rapidity of fire. Plastics and high explosives have been produced far transcending the imagination of our ancestors. As we see in the still-continuing agony of the Lebanon, rocket-propelled bombs are a commonplace, and there is not much in the carefully-guarded and "sophisticated" armouries of the major powers that has not become available to the unofficial "militias" and "guerrillas". Each member of these has destructive and lethal powers at his disposal which were undreamed of in Victorian and earlier times.

Today, people who agree with his aims call him a "freedom-fighter"; people who happen to disagree with them call him a "terrorist". What a world we have to live in! History has moved through a complete cycle, and alas once again "every man is potentially a soldier". Soon it will be every woman too. The clash of trained and organised armies was bad enough, calculated to lead to many heart-searchings of guilt and pacifism, yet this is far worse. But it is indeed our plight today, and my belief is that nothing short of universal, social, political and international justice will relieve it.

The essays in this book were first written at a time when the Soviet Union still "had the dew on it", and everyone looked to it for progressive policies. I remember a book-title of those days, "Moscow has a Plan", which suggested that nobody else had. But now it is becoming fairly evident that there are some techniques and practices which socialism could usefully take from capitalism, and there is a general move — in China, Hungary, Yugoslavia, etc. — towards a "mixed economy". That might well be the pattern of the future.

Yet the biggest difference between the days when these essays were written and now concerns science itself. All my friends and fellow-writers believed that applied science was essentially beneficial to mankind; and our objection to capitalism was that it prevented this process taking place. I remember that the burning of surplus coffee in Latin America was always quoted as an example. Yet over-production is still a menace to the capitalist world, and today we have butter-mountains and wine-lakes, while at the same time millions of human beings are starving in Africa. Something is wrong somewhere.

But now people are not at all sure that applied science is necessarily beneficial to mankind. Above all, atomic weapons, but also atomic power, toxic waste, acid rain, computerisation, the so-called "information revolution", and then genetic engineering, with its dreadful potential dangers, or the miracles of organic chemistry and all they can do, are examples of what frightens people. It may well be they are quite right to be frightened, when one recollects the escape of toxic gases from chemical factories, as happened recently in India. At the present time, the capitalist world is forging ahead fastest with these technologies, and the People's Democracies are always trying hard to get them too. Sometimes they succeed. They may well have a better chance of controlling them.

Yet unrestricted capitalism, unlike socialism, is based on greed and acquisitiveness, of which privatisation is only one aspect, and this is why in the end it cannot win. It goes against the whole trend of social evolution from the beginning of time. The tendency to do evil may be innate in man, hence the doctrine of original sin. But he and she are also under the power of the moral imperative, and that has been the driving-force of social evolution. This was why I said in the book that socialism is inevitable, as well as good and right. The Kingdom of God on Earth will assuredly come, but the time that it will take for this to happen will be the result of our own actions. "The saints under the altar cry 'O Lord, how long, how long?'" The ultimate outcome is sure, but the time which the process takes will depend upon us.

The work of this construction is the most sublime object to which man can devote himself. It has been going on ever since the time of primitive savagery, and over and over again, evil hands have broken down what has been laboriously built up, but mankind always starts again, and progress is real.

Seen from the standpoint of history of science, the changes which have had most impact on the world have been those which took place during the late Renaissance, and specifically the Scientific Revolution. This means the transition from ancient and medieval science, which always had an ethnic stamp, to the universalism of modern science, i.e. the mathematisation of hypotheses about Nature, allied with relentless experimentation.

Every people on earth has entered into this inheritance, and every person, irrespective of class, sex, colour or creed can use it once they are trained for it. Without it, diseases cannot be cured, and aeroplanes will not fly. But when they fly, it doesn't matter in the least whether the pilot is an Indonesian, the co-pilot a Swede, the navigator a Nigerian and the flight engineer a Mexican — if they are well-trained technologists, we all feel absolutely safe, and we can be assured of happy landings.

Of course, modern science was built upon the foundations of ancient and medieval sciences. I have already referred to the rocket, and indeed it was about 1150 A.D. that Chinese military engineers, whose names are not known to us, produced the first gunpowder rockets. The successful use of these for war by the Indians of the 18th century led to their development in the West, which applied modern scientific knowledge to them, and eventually those almighty rocket engines which alone, of all vehicles, can carry man and his artifacts far out into space, making it possible to visit (perhaps ultimately to inhabit) the moon, the other planets and the stars.

But by an extraordinary paradox, these same great rocket motors can be used to carry nuclear warheads right across the world and to bring about such destruction of mankind and human culture as Hiroshima and Nagasaki were but foretastes of. What was true in the case of fire itself, discovered aeons ago by primitive man, still remains true of these dreadful powers today. Everthing depends upon what you do with it. Good men can use it to warm their dear ones and to cook food; evil men use it to kill or torture others.

But the Scientific Revolution did not happen by itself. It had two concomitants, the rise of capitalism and the Protestant Reformation. This was the package deal which happened in Europe in the 17th century, and in Europe alone; and the terrifying power which these developments gave to Europeans has without doubt been the main cause of the Europocentrism which we are all busy trying to combat today.

First it gave a primacy of armaments, wherewith Europeans were able to dominate all the other peoples of the world, and inaugurate a period of imperialism and colonialism which is only now coming to an end. Modern technology, springing from modern science, was essentially responsible for this, or more truly and correctly, the deeds of evil men making use of modern technology. Yet many historians still at the present day fail to appreciate these things. Books can be written, for example, about the way in which the European full-rigged ship quickly outclassed the Chinese ocean-going junk, because of its guns and sails — yet in the index there will be no entry at all, either for "capitalism" or for "metallurgy" or "technology".

And so in the course of time, the physicists began to explore the sub-atomic world, finding there indeed a state of affairs vastly different from the ordinary Newtonian world which is appropriate for beings our size. The powers of the nature of matter, hidden in the sub-atomic depths, were studied and harnessed for human use.

But at the same time, just as fire itself could be used either for good or evil, the nuclear atomic power was applied for the purpose of nuclear weapons infinitely more destructive and dangerous than any weapon known to earlier generations. And again there is a paradox, namely that all this was done by those who were fighting the fascism of the Nazis, in a good cause, but with what appalling effects.

On the one hand, it is a good thing that modern science should be universal, and that all men can use it everywhere in the world once they are trained to do so; but it is a very bad thing that nuclear weapons should spread everywhere in this imperfect world of nation-states, often in the grip of megalomaniacs or fanatical fundamentalists, to say nothing of the activities of small groups of people blindly believing in the justice of their causes.

Humanity today is in urgent need of learning one fundamental lesson. In the *Kuan Yin Tzu* book, a Taoist text written in the 9th century A.D., the writer is talking about all sorts of wonders of nature and art, including the making of exceedingly sharp swords. "Only those who have the Tao", he says, "will be able to perform such actions and better still, not perform them, though capable of performing them." The very survival of humanity is at stake, if it will not learn that while we may well know, we should not necessarily apply our knowledge. Knowledge cannot be refused, it can never be bad in itself, but man must refrain from practising all that it is possible for him to practice.

The "technological imperative" which dictates that everything possible must be done, whether it is in the interests of men and women that it should be done or not, must absolutely be resisted. Only so can a world fit for human beings to live in be created in the future. One wonders how humanity will learn this? "They have Moses, Marx and the prophets. Let them hear them! Alas, even if one rose from the dead, they will not believe." Humanity has had its great seers and prophets — Jesus Chrismatos, Gautama Buddha, Muhammed, but their messages still fall on deaf ears.

Do we need perhaps some new prophet, speaking the language of the nuclear age? Every age can produce prophets like him who brought the whole of a people as numerous as one-sixth of humanity out of the desert into the promised land. And that was in our own time. I mean, of course, Mao Tsê-Tung. Whoever it is will have to say that humanity must learn to know, yet not to use. That seems to be the essential lesson that humanity has to learn, and if he does, all convictions of superiority will merge into our common inheritance, and all mankind will be one. It is our basic faith that this will happen. And it is more than a faith, because the Kingdom of God on Earth has all the authority of the evolutionary process behind it.

Joseph Needham
September 1985

Metamorphoses of Scepticism
Introductory Essay (1941)

"He says goodbye
To much, but not to love. For loving now shall be
The close handclasp of the waters about his trusting keel,
Buoyant they make his home, and lift his heart high,
Among their marching multitude he never shall feel lonely.
Love for him no longer a soft and garden sigh
Ruffling at evening the petalled composure of the senses;
But a wind all hours and everywhere he no wise can deny."
(Chorus from *Noah and the Waters*, C. Day Lewis).

THIS book is the third of a series of collections of essays and addresses; the first, *The Sceptical Biologist*, appeared in 1929,[1] and the second, *The Great Amphibium*, appeared in 1931.[2] Even to-day there are many professional scientists who look askance at the action of a colleague who dares to speak out from time to time on general topics. In his spare time, they feel, he should occupy himself with some innocent and health-giving occupation such as golf or fishing, rather than with dubious studies in the history of science or philosophy, the development of economic structures, the ramifications of folk-lore or the language of the Aztecs and the literature of Cathay. The overt rationalisation of this feeling is that a scientific worker can hardly be thought to have sufficient intellectual energy for his scientific work unless he is careful to use none outside it—apart from the fact that general inquiries have a strangely irritating quality on those who quite honestly do prefer golf or bridge. But the real meaning of this feeling is that to enquire too curiously into the structure of the world and society and the history of society is potentially a menace to the stability of society. The innocent scientist who harbours no "dangerous thoughts" is a far more wholesome member of the community (from the point of view of its *de facto* rulers) than the scientist who prefers to prowl. Like the various departments of some great industrial plant, into which one must not penetrate without a special pass, History and Philosophy would be presented as closed doors, if it were possible, to the scientific investigator in the pay of the bourgeoisie.

[1] (Chatto & Windus, London, 1929: Norton, New York, 1930.)
[2] (S.C.M. Press, London, 1931.)

I am glad to confess that, like T. H. Huxley, du Bois Reymond, Ernst Haeckel, and many another better man, I have always been a prowler, an explorer, among ideas. The *Sceptical Biologist* and the *Great Amphibium*[1] represented a fairly systematic position in the philosophy of science, but as would be expected in a prowler among ideas, this position has during the past ten years undergone a considerable reconstruction. The first necessity in introducing the present book is to indicate how it links on with the two former ones.

The Differentiation of the Forms of Experience.

The task with which I was mainly occupied in SB and GA was the distinctification and differentiation of the great forms of human experience; science, philosophy, religion, history and art. During the previous ten years I had often been nauseated at the confusion of them all together which is so common among superficial thinkers in pulpits and elsewhere, and I tried therefore to show how different each of them was from the others. Each seemed to lead to a characteristic world-view, incompatible with and sometimes frankly contradictory to those of the others. The proper appreciation of the world by man could not arise, I believed, from the pursuit of any one of these forms of experience by itself, but rather by the experience of all of them, though there was little or no hope of uniting them into any kind of "philosophia prima" or coherent view of the universe. The use of the word "scepticism" in a title implied, therefore, two things; first that I was sceptical of any one of the forms of experience claiming to be a royal road to our appreciation of the world in which we live, and secondly that I was sceptical of the construction of any coherent world-view which would reconcile the conflicting claims of the forms of experience together. In this sense I deliberately use the word "appreciation" instead of understanding or comprehension, for the purely intellectual would at once cut out the contributions of religion and art, forms of experience which have something in common with being in love. A man who had never been in love might give us a proposal for a world-view, but if we knew this fact about him, we should be right to think twice before accepting it. And the saying of Sir Thomas Browne, "Thus is Man that great and true Amphibium, whose nature is disposed to live, not only like other creatures in divers elements, but in divided and distinguished worlds" again emphasised my scepticism as to the possibility of a coherent world-view.

[1] Hereinafter referred to as SB and GA respectively.

I am still a sceptic, but certain processes of metamorphosis have taken place.

We may pause for a moment to note the delineation of the forms of experience. Science (as I then saw it) is abstract, dealing with statistics and avoiding the individual, aiming at the establishment of natural regularities ("scientific truth"), and quantitative, metrical, mathematical and deterministic. It is essentially classificatory, mechanical, analytical and orderly, generalising and impersonal. It has ethical and aesthetic neutrality. It fights mystery and teleology, and it gives a characteristic peace of mind, the Epicurean "ataraxia," $\dot{a}\tau a\rho a\xi i a$, akin to the early Taoist conception of "chêng ching," 正 靜. It is both rational and empirical. Religion, on the other hand, is concrete and individual, based on the sense of the holy, with which are connected the sentiments of reverence and awe. It is qualitative in feeling, opposed to measurement and analysis, "cornucopial" instead of orderly, personal rather than impersonal, and essentially irrational and alogical in spite of the cloak of rational thought based on uncertain premises which rational theologians, such as the scholastics, have sometimes succeeded in throwing about it. It naturally insists on "free-will" as against determinism. As for Philosophy, it partakes of the abstraction typical of science, but it is also interested in the individual, the qualitative, and the teleological. Unlike science, it claims to be normative, and it is not so vigorously opposed to mysteries and paradoxes. History again partakes of many of the attributes of science, but its criteria of evidence are not quite the same, and since, like philosophy, it is debarred from the performance of any experiments, its conclusions can never be tested in that way. Predictions are rather in the province of science. Artistic experience finally approximates to the religious domain, in that it is sharply separated from the rational and the expressible, but its importance to man is as great as any of the others, and there is some connection between the appreciation of beauty and the divination of the holy. Such contrasts will be found at length in SB and GA.[1]

Looking back at my efforts ten to fifteen years ago to disentangle the forms of experience, I think now that the description of science was rather too narrow, and the description of religion certainly much

---

[1] This kind of treatment, done far better than I, as a working scientist, could ever have hoped to do it, was given in R. G. Collingwood's *Speculum Mentis* (Oxford, 1924); and it is interesting that the subsequent development of this philosopher, as his *Autobiography* (Oxford, 1939) shows, has paralleled my own.

too neo-Platonic, idealistic, pietistic and other-worldly. In my anxiety to see the forms of human experience in sharpest antithesis to one another, I almost welcomed characteristics which added to their irreconcilability. I was determined to be a "divider," and if indeed the work of previous "uniters" was insufficiently well-founded, this was no mistake. "Dividing" and "uniting" may be looked upon as dialectical opposites, like continuity and discontinuity, which run through the whole history of philosophical and scientific thought, imperfect unions being always doomed to destruction by later "dividing" critics.

But much the most significant thing about my point of view at that time, as I remember it, was that I was always uncomfortable about the position of ethics. It was the one department of thought which I could never allot to one or other of the forms of experience. It seemed sometimes to belong to religion, but yet it was obviously profoundly affected by science, and in some situations in time and space, such as early Chinese Confucianism, it existed without any connections with supernaturalism. Nor was it remote from aesthetic experience, since good actions had something beautiful about them; nor could it be considered apart from history, since systems of ethics had changed with changing social conditions.

The explanation of this difficulty was at hand, however.

The Breaking in of Ethics and Politics.

During the past twenty years, the one form of experience which in SB and GA was never taken into account,[1] came and forced itself more and more upon my attention, namely politics. Ethics and politics both belong to the realm of man's social life, and my fundamental limitation had been to envisage the experiencing human being as a solitary unit, subject to the diverse forms of experience which I had been classifying. Ethics are the rules whereby men may live together in society with the utmost harmony and the best opportunities for the development of their talents in the common good. Politics are nothing but the attempt to objectify the most advanced ethics in the structure of society, to enmesh the ideal ethical relations in the real world. Hence the class struggles in every age since the origin of private property, which, though at many stages a fundamentally progressive force, at other stages radically hinders the attainment of the next step in social relations. Now so far as my thinking was concerned,

[1] There were indications of it in GA, however, cf. p. 43.

ethics and politics proved to be the cement necessary for the unification of the divergent forms of experience. The dividing process was succeeded by a uniting one, and an integrated world-view emerged from the differentiated dissected analysed system which I had made. It was bound to follow the lead of the philosophy which most consistently allows for the social background of our thought and being, and explains what is happening, and has for centuries been happening, to human society as the continuation of all biological evolution.

Perhaps it will not be a digression if at this point I lighten the way by a few autobiographical notes. I tried to keep to my own field, but politics would keep breaking in. I grew up in an extremely bourgeois household, my father a physician (before my time, an anatomist and pioneer in pathological histology) in private practice but later to be a specialist in anaesthesia; my mother a musician and composer. My father was, I believe, an extremely kind man in his practice, among which there were many working-class people, but the atmosphere of my home was saturated with every kind of bourgeois prejudice. I remember with affection, however, an incident which was the first to make me realise the community of flesh and blood which I had with working-class people, who otherwise one might have been tempted to suppose were an entirely different species of living organism. It was at the little French town of Eu, where I (then about thirteen) was travelling with my father on holiday, and as there was no "correspondance" of trains, we had to stay the night. The hotel was full, so we were accommodated in a neighbouring railwayman's cottage, the simple homely welcome of whose family I never forgot. Such little incidents have untold consequences, and about 1917 in walks with my father I would always argue in favour of socialism against his incurably pessimistic views of human nature, so well suited to the retention of power and privilege by the class which at any given time possesses it.

I remember, too, the first mention I ever heard of bolshevism. I was on my bicycle, while my best school friend, F—. C—. (later Professor of Architecture at M'Gill University) walked along beside me. It was the summer of 1918. We had roving minds; he did a good deal of sculpture, and we corresponded in a variety of codes, in Leonardo's mirror-writing, and so on. He was saying, "The point about bolshevism is, that so far, every ruling class has oppressed everybody else, so now the idea is to take the very dregs of the population and make them the ruling class, so that they can take a turn at oppressing everybody else." I remember saying I thought it a very *peculiar* theory. I still smile at this, but it was something to be discussing it at all; I believe most of the boys with whom I was at school did not begin to think about it till many years afterwards, if ever.

Later on, as an undergraduate, my interests were largely philosophical and theological, and I was so unpolitical that I never joined the Union Society, on account of a disinclination to be drawn into political debates. But in so far as I was anything, I was vaguely progressive, and remember supporting with great enthusiasm (and owing to my upbringing no small inside knowledge) the proposal for a State Medical Service at a debate at the Junior Acton Club, an organisation which united mostly Caius and Emmanuel men but now long since defunct. An influence of considerable importance, after my marriage, was that of our friend the biochemist L—. R—., whose eastern and central European outlook on political questions revolutionised ours. We had got to know him during a period of work at the Roscoff Marine Biological Station, and I remember discussions on the roof, in 1925, while the sun sank over the Ile de Batz, and again in our house in Cambridge later, which were very formative for us.

The process of socialisation of my outlook, however, really began with the General Strike in 1926 and was completed by the rise to power of Hitlerite fascism in 1933. In the general strike I was on the wrong side, and helped in the running of the railway as a volunteer, explaining my action to socialist friends as a straightforward support of constitutionally elected government, a government which I had certainly voted against in the preceding elections. I so far acted up to my beliefs in this way that at the conclusion of the strike, when the railway company wanted us to remain at our posts in order that their officials could conduct a victimisation process against the returning railwaymen, I spoke against this at a meeting of the volunteers, pointing out that we had no quarrel with the railwaymen, and so helped to ensure that most of the volunteers left without delay. There can be no doubt that these events supplied the most powerful stimulus I had ever had towards reading along sociological and political lines. I carried a little old torn copy of the *Tale of Two Cities* on the footplate, but in the evenings and afterwards I read Shaw's *Guide to Capitalism and Socialism* and went on from that to heavier material. In this way I came to the belief I now hold, that, in a sense, in any doubtful case, "the people are *never* wrong"; through all the ages of oppression since the first beginnings of private property men have been struggling for political freedom, and everything which assists this struggle is right. The later years of work in the Labour movement of my wife and myself, including a long membership of the Cambridge Trades Council, all originated from this time.

By 1933 the general movement among all kinds of intellectual workers in England towards political activity was becoming widespread. In the Universities it was evident in the much increased interest of the undergraduate in political thought, and the concomitant decline in organised large-scale jokes and similar amusing, if childish, exhibitions. A certain loss of lightheartedness was unavoidable. Even in the world of science, our witty periodical "Brighter Biochemistry," which in its day had had a wide reputation and had been very worth while, died after some seven years of life about this time. The situation was, in fact, getting beyond a joke, and the shadow of the Second World War was upon us.

The Creativeness of Contradictions.

I must now turn to sketch the general world-view at which I arrived after I began to realise man as a social being, and in which the various forms of experience which I had so carefully distinguished found their place. Essentially what I had unearthed from these sharp antitheses was a series of contradictions.[1] Now contradictions and deadlocks are far from being a calamity in practical thought; they are only so in formal logic. In practice they find themselves overcome by syntheses at higher levels. And the idea that "the principle of contradiction is only valid for our reason" is an ancient one; it is found in the mystical writer Dionysius the Areopagite, and again when the middle ages were giving place to the modern era, in Nicholas of Cusa, and in Giordano Bruno.[2] For example, in Master Eckhardt we find the following, easily decipherable through its archaic German:[3] "Waz ist widersatzunge? Lieb unde leit, wiz unde swarz, daz hat widersatzunge, unde die enblibet in wesenne niht. Swenne diu Seele kumt in daz lieht der Vernünftekeit, so weiz si niht widersatzunge."

All these thinkers placed the reconciliation of contradictions, however, in the realm of the divine, and did not consider it as attainable by man in his earthly state. This was the position I too adopted in my earlier thinking,[4] quoting Robert Boyle: "When we die God

[1] Cf. p. 182.
[2] See "The Cosmology of Giordano Bruno" by Dorothea W. Singer, Isis, 1941, **33**, 187, and her forthcoming book on this remarkable man.
[3] See *Nicholas of Cusa* by H. Bett (London, 1932), p. 127.   [4] Cf. SB, p. 225.

will enlarge our faculties so as to enable us to gaze without being dazzled upon those sublime and radiant truths whose *harmony* as well as splendour we shall then be qualified to discover, and consequently with transports to admire." It was a thought not infrequently met with in the 17th century; thus Sir Thomas Browne said, "There is yet another conceit that hath made me sometimes shut my books, which tells me it is a vanity to waste our days in the blind pursuit of knowledge; it is but attending a little longer, and we shall enjoy that by instinct and infusion, which we endeavour at here by labour and inquisition. It is better to sit down in a modest ignorance, and rest contented with the natural blessing of our own reasons, than buy the uncertain knowledge of this life with sweat and vexation, which Death gives every fool gratis, and is an accessory of our glorification."[1]

But it was left to Karl Marx and Frederick Engels in the last century, building on the dialectic process of the idealist philosopher Hegel, but profoundly influenced through Darwin[2] by the new understanding of evolution which was then dawning on men, to take the revolutionary step of placing the resolution of contradictions within the historical and pre-historical process itself. Contradictions are not resolved only in heaven; they are resolved right here, some in the past, some now, and some in time to come. This is the dialectical materialist way of expounding cosmic development, biological evolution, and social evolution, including all history.

However fascists may laugh, there is something divine about a committee.[3] One member makes a proposal, another points out that for one reason or another it will not do; there are further proposals and counter-proposals, and out of this strife of theses and antitheses, the synthesis, or final decision, is born, only to become itself in due time another thesis, again unsatisfactory, and to be subsumed in further higher syntheses by later meetings or higher committees. This is the dialectical process.[4] Marx and Engels were bold enough to

---

[1] *Religio Medici*, I. 9.
[2] Marx wanted to dedicate part of *Capital* to Darwin (cf. Psyche, 1931, **12**, 7).
[3] This is why I like that famous dictum of Conrad Noel's that the universe itself is ruled by a committee. The Christian doctrine of the Trinity, "neither afore nor after other," "without any difference or inequality," upheld by Athanasius, would thus have been a democratic doctrine as against the totalitarian monotheism of the imperial and Arian party. The same thought occurs in Feuerbach's *Essence of Christianity* (London, 1854), pp. 66 and 288). "The mystery of the Trinity," he says, "is the mystery of participated, social, life, the mystery of I and You," and he gives patristic sources in support of it.
[4] The best introduction to dialectical materialist philosophy known to me is that of R. Maublanc; *La Philosophie du Marxisme* (Paris, 1935); it is a pity there is no English

assert that it happens actually in evolving nature itself, and that the undoubted fact that it happens in our thought about nature is because we and our thought are a part of nature. We cannot consider nature otherwise than as a series of levels of organisation, a series of dialectical syntheses. From ultimate physical particle to atom, from atom to molecule, from molecule to colloidal aggregate, from aggregate to living cell, from cell to organ, from organ to body, from animal body to social association, the series of organisational levels is complete. Nothing but energy (as we now call matter and motion) and the levels of organisation (or the stabilised dialectical syntheses) at different levels have been required for the building of our world. The consequences of this point of view are boundless. Social evolution is continuous with biological evolution, and the higher stages of social organisation, embodied in advanced ethics and in socialism, are not a pious hope based on optimistic ideas about human nature, but the necessary consequence of all foregoing evolution. We are in the midst of the dialectical process, which is not likely to stop at the bidding of those who sit, like Canute, with their feet in the water forbidding the flood of the tide.

I shall not emphasise these consequences here, since in the other essays in this book they are further dealt with. I shall only point out that they fundamentally affect our attitude to time. Hence the title of the present book. As Auden says:—

"And the poor in their fireless lodgings, dropping the sheets
  Of the evening paper; 'Our day is our loss, O show us
    History the operator, the
  Organiser, Time the refreshing river!"

Sir Thomas Browne was wrong; "the great Mutations of the world" are not all yet acted, and time will not be too short for the development of human society that is to come. Contrary presentations, of course, spring to the mind:—

"Time, like an ever-rolling stream
  Bears all its sons away,
They fly forgotten, as a dream
  Dies at the opening day."

translation. M. J. Adler in his *Dialectic* (London, 1927, p. 10) makes the interesting point that dialectical thinking is essentially social thinking within the one mind, since it involves a conflict of opposites, which deduction and induction do not. See also the article of A. M. Dunham on the concept of "tension" in psychology and logic (Psychiatry, 1938, **1**, 79).

as we used to sing as children. But now we have no further interest in this over-individualistic type of christianity; those who are prepared to "work illegally, and be anonymous" that the Kingdom may come, find that this attitude has no meaning for them. Nor will they be depressed by the cultured hopelessness of T. S. Eliot in "East Coker":—

> "And so each venture
> Is a new beginning, a raid on the inarticulate
> With shabby equipment always deteriorating
> In the general mess of imprecision of feeling,
> Undisciplined squads of emotion. And what there is to conquer
> By strength and submission, has already been discovered
> Once or twice, or several times, by men whom one cannot hope
> To emulate—but there is no competition—
> There is only the fight to recover what has been lost
> And found and lost again and again: and now, under conditions
> That seem unpropitious. But perhaps neither gain nor loss.
> For us, there is only the trying. The rest is not our business."

Again, this is too individualist. Time is for all men a refreshing river; not merely a perpetual recurrence of opportunities for individual souls to scale the heights of mystical experience or to produce great artistic achievement or to break free from the wheel of things or to attain perfect nonactivity, or whatever metaphor of individual perfection you happen to like. The historical process is the organiser of the City of God, and those who work at its building are (in the ancient language) the ministers of the Most High. Of course there have been setbacks innumerable, but the curve of the development of human society pursues its way across the graph of history with statistical certainty, heeding neither the many points which fall beneath it, nor those many more hopeful ones which lie above its average sweep.

Instances of dialectical development in scientific thought are so numerous that a few moments' thought provides an embarrassingly large selection. All science progresses by new hypotheses which combine in a synthetic way, not by mere compromise, the truest points in the preceding hypotheses. Deadlocks are thus overcome. Thus in embryology we know now that both egg and sperm are essential contributions to generation from the two parents, but in the 18th century and well into the 19th this was not understood. The

Ovists believed that mammals developed from the egg alone, and that the spermatozoa were adventitious worm-like organisms, perhaps parasites, as the termination "-zoa" of their modern name still indicates to-day. The Animalculists, on the other hand, believed that the animal originated from a spermatozoon only, and that the egg was if anything only a kind of box in which it could develop. The phenomena of inheritance alone should have sufficed to indicate that this deadlock was an absurdity, but it took nearly two centuries before the functions of egg and spermatozoon were understood, and the contradiction was resolved. In the later history of embryology we find a similar contradiction. When the study of the fate of parts of the egg began, it was at first believed that all development is mosaic, i.e. that any injury to the egg at the beginning of development is reflected in a corresponding injury in the finished embryo. Then came the discovery that in some eggs, at any rate, regulative development can occur, i.e. that from a half-egg, or even a quarter-egg, a normal embryo, though small, can be formed. These facts were made the basis of vitalistic theories, but the sharp contradiction was at length resolved by the finding that the eggs of some species are mosaic and that those of others are regulative, the only difference between them being the exact time at which the determination of the fates of the various parts occurs.

The same syntheses of contradictions are going on in biochemistry. Some thirty years ago lactic acid was thought to be the causative agent in the contraction of muscle. After the discovery of phosphagen, this substance in turn was regarded as the most important. We know now that neither of these substances is connected with the final process whereby energy is transferred from chemical processes to the muscle fibre but another substance altogether, adenylpyrophosphate. Phosphagen is one of the substances involved in the cycles of phosphorylation by which energy is transferred, while the lactic acid is simply the waste-product of the breakdown of glycogen whereby the chemical energy is provided. Or again, adrenalin has long been known to contract blood-vessels and also raise the blood sugar by mobilising liver glycogen. This latter action was thought to be due to the constriction of the liver's small blood-vessels, with consequent anoxaemia and asphyxia. But it was then found that even in well-oxygenated liver cells, adrenalin causes a breakdown of glycogen. This contradiction was subsequently resolved by the further finding that adrenalin intervenes in the oxidative synthesis of glycogen in the cell.

A particularly neat example of the resolution of a dialectical contradiction is that of nucleic acid synthesis by developing eggs. It was first found that during the development of sea-urchin eggs, a large quantity of histologically recognisable nuclein is formed for the increasing number of nuclei. But it was then found that there was no change whatever in the nuclein phosphorus or the nucleic acid nitrogen during development. Here was a flat contradiction, but it was resolved by workers who showed that there are two kinds of nucleic acid, and that during development one of them, situated in the cytoplasm, is transformed into the other kind situated in the new nuclei.

In the field of wider ideas, there is a convincing sense in which one may say that the long-debated controversy between biological mechanists and vitalists (much discussed in SB and GA) was a dialectical deadlock which a judicious organicism has resolved.[1] The mechanists, enamoured of over-simplified physico-chemical explanations of biological processes, which they regarded, quite rightly, as heuristically valuable, maintained that all biological processes were fully explicable in terms appropriate to the sciences of physics and chemistry. The vitalists, always eager to safeguard objective complexity (and at the same time to keep the world safe for animism), maintained that vital phenomena would always escape physico-chemical analysis. This deadlock, which in various forms had run through the whole history of human thought, was overcome when it was realised that every level of organisation has its own regularities and principles, not reducible to those appropriate to lower levels of organisation, nor applicable to higher levels, but at the same time in no way inscrutable or immune from scientific analysis and comprehension. Thus the rules which are followed in experimental morphology or genetics are perfectly valid in their own right, but comprehension will never be complete until what is going on at the other levels, both above and below, is analysed and compared with the level in question.[2] Biological organisation is the basic problem of biology; it is not an axiom from which biology must start.

So in the same way, we may perhaps consider dialectical materialism itself as the synthesis of the age-old contradiction between meta-

---

[1] This idea was put forward in a paper of mine in 1928, Quart. Rev. Biol., **3**, 80.

[2] M. J. Adler (*Dialectic*, London, 1927, p. 164 ff.) suggests that the natural connection between dialectical and organicistic thought is simply that entities in opposition are likely to be parts (on one level) of which the whole, the synthesis, occupies the next higher level.

physical idealism and materialism.[1] Idealism did justice to the highest manifestations of human social activity, but was absolutely incapable of doing justice to that real objective world on which science must insist. Materialism provided a world congruent with scientific activity, but, since it lacked the secondary qualities, a world so "grey and cimmerian" (in Goethe's phrase) that only the stoutest-hearted or the thickest-headed could accept it as an account of that real world in which some place had to be found for human values. By their fight against both classical idealism and mechanical materialism, and by their insistence on the successive dialectical levels in nature, the highest of them including all man's highest experiences, Marx and Engels overcame this contradiction for the first time in history. Engels pointed out three limitations in former mechanical materialism; first that it was mechanical "in the sense that it believed in the exclusive application of the standards of mechanics to processes of a chemical and organic nature"; second, that it was anti-dialectical, i.e. that it did not allow for the ever-shifting boundaries in nature; and third, that it admitted of the preservation of idealism "up above" in the realm of the social sciences.[2]

In historical events the dialectical process is readily seen.[3] The English civil war in the 17th century provides a particularly striking example. The whole feudal systems of ideas, represented by the King, the aristocracy and the Anglican bishops came into sharp opposition against the rising middle class led by the smaller country gentry, the merchants and the City of London. Feudal royalism found its antithesis in the radical puritan republicanism of the Commonwealth period. But the time was not ripe for the ideas of the Levellers and Independents, and the Restoration was a dialectical synthesis in which a constitutional monarchy, or one which was bound to end as such, combined with a triumphant Parliament controlled by the middle class whose interests the civil war had made secure. The protestant coup d'ètat of 1688 and the Reform Act of 1832 simply completed

[1] Cf. Lenin *On Dialectics* in Works, **11**, p. 84. Or one might say that Thomas Aquinas made a dialectical synthesis of the points of view represented in earlier scholastic thought by Abelard and William of Champeaux (M. J. Adler, *Dialectics*, London, 1927, p. 72). Clement of Alexandria actually did say, in the *Stromata*, that Christianity was the synthesis of Greek and Jew.
[2] Cf. Lenin's *Materialism and Empirio-Criticism* in Works, **11**. p. 297.
[3] As for example by the non-marxist historian H. Butterfield in his interesting book *The Whig Interpretation of History* (London, 1931), where he combats the moralising attitude to historical conflicts, showing that each side stood for some elements which were embodied in the subsequent synthesis.

the process. The French Revolution shows a very similar development. The feudal monarchy was opposed by the revolutionary Jacobins, but the eventual outcome was the rule of the post-Napoleonic bourgeoisie. This dialectical process is the explanation of the feature so characteristic of revolutions, that they move (in common parlance), two steps forward and one step back.[1]

A Reconsideration of Beliefs.

In the light of what has now been said, it may be of interest to reconsider some of the points of view in my two previous books of essays. In the first place the attack on vitalism in all its forms[2] was abundantly justified; there is no place whatever in biology for traces of animism, but we must seek to understand the biological level side by side with the physico-chemical level and the psychological level. In saying that living things differ from dead things in degree and not in kind, and are, as it were, extrapolations from the inorganic,[3] I was explicitly adumbrating the scheme of successive levels of complexity and organisation. The essay "Lucretius Redivivus"[4] was sound in that it emphasised a coming connection between chemical science and mental science, and I look back with pleasure on my enthusiasm for Epicurus and Lucretius, from which I have never seen any reason to depart, and which has since been publicly shared by others in some extremely valuable books.[5] My dislike of fixed boundary-lines in nature[6] was, I found, in agreement with what Engels says on the question in that great but unfortunately named work the *Anti-Dühring*."[7] At that time, much influenced by Lotze, whom I still consider a remarkable thinker, I emphasised that mechanism was to be considered applicable everywhere, but final nowhere.[8] This was an inadequate way of saying that the scientific method is applicable at all levels but that mechanical explanations are inadequate to deal with the phenomena of organisms. There was a similar confusion between "scientific naturalism" and "mechanical materialism"; I often wrote the former when, as I think now, I should have written the

---

[1] The examples given above show the dialectical process at work in the history of human society and of scientific thought. But it is embodied also in non-human evolution, see p. 190.
[2] SB, pp. 89 ff.; GA, pp. 95 ff.   [3] SB, p. 247.   [4] SB, p. 133.
[5] Such as B. Farrington's *Science and Politics in the Ancient World* (London, 1939).
[6] SB, p. 16.
[7] Also in *Dialectics of Nature* (Gesamtausgabe edition, Moscow, 1935, p. 629).
[8] SB, pp. 28, 136.

latter.[1] But this was because I could at that time see no way of including the highest phases of organisation within the realm of nature without subjecting them to the distortion which mechanical materialism put upon them, a distortion which involved the characteristic denial I could never bring myself to make, the denial of the validity of one or other of the forms of human experience. Religion and art, for instance, are perfectly valid forms, though they certainly do not mean what their interpreters and experiencers have often thought they meant. Mechanism, then, according to my view, was applicable everywhere but final nowhere, and in biology this led to a position of methodological mechanism for which, in opposition to the neo-vitalists, I used the term neo-mechanism.[2] This was a way of acknowledging the complexity of the high organic levels without giving up the scientific method. It was also a way of acknowledging the imperfection and relativity of the scientific formulations so far attained. It went with, though it did not necessitate, the idea that the other forms of experience, such as religion, philosophy, history and art, were alternative, also imperfect, ways of apprehending the world in which we live. Owing to my ignorance of any form of naturalism other than mechanical materialism, my discussion of biological organicism was somewhat vitiated. While sympathetic to organicism, I assumed that biological organisation could not be investigated scientifically, and must therefore remain a concept of purely philosophical order.[3] Later I was able to revise this view thoroughly, and to show that on the contrary, organising relations *are* open to investigation.[4] It is precisely the organisation of the various levels that constitutes their special quality and gives rise to their special forms of behaviour.

I think it is fair to say that my presentation of the main metaphysical issue was never idealist.[5] But I was much influenced by the trend of thought originating with Ernst Mach, and agreed that the procedures of science do not give us a picture of the external world as it really *is*. While maintaining the real existence of the world (of matter) prior to ourselves, I described many features of the scientific method which suggest that our knowledge of it comes to us in distorted form. However much one may think this distortion amounts to, as long as one admits a knowable objective basis of our experience, one remains a materialist. My ideas on this subject were not cleared up until I read Lenin's book on the Machians, *Materialism and Empirio-*

[1] SB, p. 242; GA, p. 101.   [2] SB, p. 38.   [3] SB, pp. 83, 84.
[4] In my book *Order and Life* (Yale and Cambridge, 1936).   [5] SB, p. 26.

Criticism. Lenin showed that the Machians were really disguised philosophical idealists, not merely affirming the existence of some distortion in our apprehension of the external world, but tending to deny its very existence or to make all science the purely subjective study of an unknowable noumenon. He elucidated the confusion which the Machians had introduced between the objective-subjective antithesis on the one hand, and the absolute-relative antithesis on the other. Scientific truth is certainly relative, since all formulations are imperfect, though approaching, perhaps asymptotically, to the truest possible account of the regularities of external nature; but it is equally certainly objective, in that it deals with real external events, even through the "optic glasses" of our human limitations.[1] Relativity of scientific truth is taken care of in dialectical materialism, since it is constantly approximating dialectically to truth; but subjectivism, if it be pure, is nothing else than metaphysical idealism in disguise.

The seemingly endless process of the improvement of human knowledge was very much in my mind when the essays of SB and GA were written.[2] Because of my doubt whether any "philosophia prima" could ever be found to reconcile the different forms of human experience, I emphasised all the more the practice of activities themselves, never doubting that each was undergoing progressive advance and refinement. Hence the quotation from the Nicomachean Ethics of Aristotle with which SB was headed: "The greatest good which man can know is the active exercise of the spirit in conformity with virtue." There is a parallel to this in the seemingly endless process of the improvement of human society. Those whose conscious or unconscious interest it is to minimise the achievements of social evolution generally choose one of two ways of attack; saying either that the art and thought of the ancients has had no equal since, or that no matter what changes the further development of human society may bring, it will never approach that perfection which has long been imaginable. The first point of view is the product of a distorted time-scale. The second is the most superficial of all pessimisms. Whether or not human suffering will always exist, it is always our duty to work to decrease it, and the course of social evolution hitherto gives us every confidence for the future. Social, as well as scientific, ethos is summed up in the words of G. E. Lessing (a favourite passage of the great

---

[1] Cf. Lenin, *Materialism and Empirio-Criticism*, pp. 185, 199, 363.
[2] SB, pp. 7, 33 ff.

physicist, Max Planck): "Not the possession of truth, but the effort in struggling to attain it, brings joy to the searcher."

The Individualist Fallacy.

All the worst deficiencies of SB and GA, as I now see it, were due to thinking of man solely as an individual, with various different facets, or windows, out of which, as in some observatory or conningtower, he could look. Immediately one begins to consider man under his social aspect, the germs of a unified world-view appear.[1] As I have said, the position of ethics was, for this very reason, always obscure to me. Ethics are the rules whereby man may live in social harmony; Confucius discussed *li* 禮 and *Jen* 仁 and *i* 義 in this sense and without any supernaturalism thousands of year ago. They are being discussed to-day in the same spirit.[2] Such rules perhaps correspond to the valency bonds and other forces which hold particles together at the molecular and sub-molecular levels. Politics is only practical mass ethics, some men seeking to perpetuate private possession of the goods of life, others seeking to distribute them as widely as may be; some men seeking for reasons why nothing should ever change, others seeking for true knowledge of the nature of man and how the natural needs can be satisfied. Hence for civilised man, in whom the numinous, the sense of the holy, is irrevocably attached to ethical ideas, religion too becomes a bond playing its part in the coherence of high social organisation. When Society itself has been sanctified by the full incorporation in it of the principles of justice, love and comradeship, religion is destined to pass without loss into social emotion as such. When oppression has been removed, religion as the cry of the oppressed creature will cease to exist, but the sense of the holy, one of man's most fundamental forms of experience, will never disappear. We can already see a similar transformation taking place in poetry, where the most moving implications can be conveyed

[1] Long before, Feuerbach had passed through precisely the same intellectual process (*Works*, pp. 343, 344). And cf. Bukharin's words, which I read long afterwards, "The philosophical 'subject' is not an isolated human atom, but 'social' man." (Marx Memorial Volume, Academy of Sciences, Moscow, 1933, Eng. tr. *Marxism and Modern Thought*, London, 1936, p. 13.) But there is no need to go so far afield; no one appreciated these things better than the great English psychologist, Henry Maudsley (cf. his *Body and Will*, London, 1883, pp. 44 and 157).

[2] See the discussion which followed a paper of C. H. Waddington's in Nature, 1941, **148**, p. 270 ff. with contributions by E. W. Barnes, W. R. Matthews, W. G. de Burgh, A. D. Ritchie, Julian Huxley and others; see also Proc. Aristot. Soc. 1942, and *Science and Ethics* (Allen & Unwin, London, 1942).

by the poets in the simplest common words. The religious mysticism of a Donne or a Crashaw; the cosmic pantheism of a Wordsworth; have given place to the social emotion of such poets as Auden, Spender, Day Lewis, and such prose writers as Warner and Upward.

There is a story by that subtlest of writers, V. S. Pritchett, which well exemplifies this.[1] It is a love story and it concerns a commercial traveller and an undertaker's daughter, but the plain ordinariness of the conversation masks the deepest feeling and the most devastatingly subtle irony. This mingling of profound emotion with surface banality strikes an authentic note of what must come to be in future socialist society.[2] Auden has supplied a perfect allegory of it in his poem:

> "To settle in this village of the heart,
> My darling, can you bear it? True, the hall
> With its yews and famous dovecots is still there
> Just as in childhood, but the grand old couple
> Who loved us all so equally are dead;
> And now it is a licensed house for tourists,
> None too particular. One of the new
> Trunk roads passes the very door already,
> And the thin cafés spring up overnight.
> The sham ornamentation, the strident swimming pool,
> The identical and townee smartness,
> Will you really see as home, and not depend
> For comfort on the chance, the sly encounter
> With the irresponsible beauty of the stranger?
> O can you see precisely in our gaucheness
> The neighbour's strongest wish, to serve and love?"

Pritchett and Auden and all our best writers are warning us not to be put off by what we may feel are the vulgar externals of modern life; not to retire into fantasies and escape-holes; the reality of human comradeship is as powerful as ever. The lorry-drivers on those roads have T.U. cards in their pockets and their talk in the thin cafés is far from fantastic. Those young men and women in the strident swimming pool are members, perhaps, of L.L.Y. or Y.C.L. The

---

[1] "Sense of Humour" in New Writing, 1936, **2**, 16, and in *You make your own Life* (London, 1938).
[2] Was this not the meaning of that great saying of Yeats—"Think the thoughts of a wise man, but speak the common language of the people."

beauty of these people is not strange, not irresponsible, but pregnant with the beauty of the new world-order.[1]

Perhaps a word about modern English poetry might be interjected here. The writings collected in the present book go to show how a scientist deeply responded to the work of poets contemporary with him during the period between the two world wars. He could not refrain from quoting them because they embodied all the elements of his own world view, the evolutionary background, the materialist view of human history, the task of Eros in social progress, the revolutionary belief in the future world of justice and comradeship. W. B. Yeats,[2] when discussing the achievements of the "New Country" School and their successors, remarked that certain technical factors such as assonance and sprung rhythm, had permitted at last the inclusion of necessary scientific words into poetry. Scientific socialism could then bring science and poetry together.

It is essential, therefore, to view all the forms of human experience in a social context. But I was profoundly sceptical of the right of any one of the forms of experience to have the last word about the world in which we live. To-day I feel more confirmed in this scepticism than ever. A concentration on scientific experience alone gives you the individualistic researcher, inapt for team-work and bent on priority, the easy prey of all the reactionary social forces tending to

---

[1] Compare with this a notable passage from George Orwell:
"The place to look for the germs of the future England is in the light-industry areas and along the arterial roads. In Slough, Dagenham, Barnet, Letchworth, Hayes—everywhere indeed on the outskirts of great towns—the old pattern is gradually changing into something new. In those vast new wildernesses of glass and brick the sharp distinctions of the older kind of town, with its slums and mansions, or the country, with its manor houses and squalid cottages, no longer exist. There are wide gradations of income, but it is the same kind of life that is being lived at different levels, in labour-saving flats or council houses, along the concrete roads, and in the naked democracy of the swimming pools. It is rather a restless, cultureless life, centreing round tinned food, *Picture Post*, the radio and the internal combustion engine. It is a civilisation in which children grow up with an intimate knowledge of magnetos and in complete ignorance of the Bible. To that civilisation belong the people who are most at home in, and most definitely of, the modern world; the technicians and the higher-paid skilled workers, the airmen and their mechanics, the radio experts, film producers, popular journalists and industrial chemists. They are the indeterminate stratum at which the older class distinctions are beginning to break down."—(*The Lion and the Unicorn*, London, 1941, p. 54.)

We are reminded of a great scholar venturing to stand up against one of Inge's diatribes against modern life—"I do not regard it as absurd," said Rashdall, "to contend that there is value even in the life of East and West Ham" (*Ideas & Ideals*, 1928, p. 85).

[2] In his introduction to the *Oxford Book of Modern English Poetry*.

make scientists the passive instruments of class domination. In looking round among one's colleagues it has been consistently evident that those with the narrowest specialist interests tend to be politically the most reactionary. Without history, the scientist will know nothing of social evolution, of the origin and progress of human society, of the laws of change and of the direction in which further progress is likely to take place. Without philosophy he can have no basic world-view, and may fall into all kinds of fantasies—for successful scientific work is compatible with anything from Roman Catholicism, as in the case of Pasteur, to Sandemanism as in the case of Faraday. It would be presumptuous in the case of such men to think that their scientific work would have been better if they had had better philosophies, but the majority of scientists are not of their calibre, and for these it is surely true that the better their philosophy the better their scientific work is likely to be. Without religion, the scientist will know little of comradeship with the mass of men, he will remain isolated from them in intellectual pride, and incapable of that humility which made Huxley give of his best to working-class audiences in "Mechanics' Institutes" or Timiriazev and Sechenov lecture illegally to Russian working men. Only by recognising where the numinous really lies will he be able to take his part in the great Labour movement. As for the absence of aesthetic appreciation, one need not describe the kind of person he will be without that. All the forms of experience are necessary and no one of them has the last word.

It would be tedious to apply the same arguments to other sorts of men. Thus the historian without science will become a donnish "period-prisoner," and without philosophy a pedantic purveyor of meaningless facts. The religious man, without political understanding, will reduce ethics to relations between individuals and will sink into the false and vicious religion of other-worldly pietism. But even should he avoid this, he would become, without some admixture of science and philosophy, a pure social revolutionary, a Utopian lacking all solid background for his faith. Everyone can apply these principles to their own cases. As old Comenius said:

> "Can any man be a good Naturalist, that is not seene in the Metaphysicks? Or a good Moralist, who is not a Naturalist? Or a Logician, who is ignorant of reall Sciences? Or a Divine, a Lawyer, or a Physician, that is no Philosopher? Or an Oratour or Poet, who is not accomplished with them all?"
>
> (*A Reformation of Schooles*, 1642.)

And so I am a sceptic still. It was not by any means a useless task to distinguish with all possible exactness the forms of experience from one another. But the conclusion now is, not as before, that a man should exercise his soul (in Aristotle's phrase) in conformity with virtue, without the hope of unifying in any way the products of its exercises.[1] It is, to view the world as a whole and the place and course of man and of humanity in it, and to know "what he must *do* while still in his compounded body." The consideration of man and his experiences as an individual led in the end to contemplation; the consideration of social man and his experiences leads to action. No more shall we take Gautama and Plato for our guide, but rather those determined men who from Confucius to Marx were vehicles of the evolutionary process, working through them to implement the Promise occluded in the very beginning of our world.

---

[1] When writing this introductory essay, I happened to be reading that great work of scholarship, George Thomson's *Aeschylus and Athens* (London, 1941), in which he describes the anthropological origins of Greek civilisation and folk-lore, and the rise of Greek literature and culture from them. From a discussion of Ionian science, especially in Anaximander (p. 83), and Orphic mystical theology (p. 156), he suggests that "the tendency of aristocratic thought is to divide, to keep things apart" while "the tendency of popular thought is to unite." In Orphism, Love implied the reunion of what had been sundered. It would be interesting to investigate the social significance of philosophical "dividers" and "uniters" in different historical times; certainly in my own development, Thomson's correlation has been strikingly substantiated. I was unable to find any unified world-view until I took man's social life into account, a thing no aristocratic thinker would ever desire to do, unless as a reaction against democratic thinkers for specifically polemical purposes.

The Naturalness of the Spiritual World
A Reappraisement of Henry Drummond

(Based upon an Introduction to a new edition of Henry Drummond's *Natural Law in the Spiritual World*, 1939)

EVERY age has its "dividers" and its "uniters." The dividers seek to distinguish between things which they believe their predecessors have confused. They are nauseated by the facile identification of ideas and modes of experience which are by essence utterly different; they pull apart in order to understand and to clarify. The uniters, on the other hand, are always straining after some unifying hypothesis, some "philosophia prima," some all-embracing world-view or some scientific hypothesis bringing hitherto unrelated groups of facts or theories into relation. The uniters fall into over-simplification; the dividers expose their shallowness without, perhaps, being able to suggest anything better. And so goes on the eternal swing between two poles; the world-view that is certainly wrong, or at least, incomplete; and the critical scepticism that is certainly no world-view. Like the ideas of continuity and discontinuity in fundamental physical theory, these poles of philosophy and criticism provide perhaps the necessary contradictions out of which the advances of human understanding are for ever being born.

Henry Drummond was certainly one of the uniters. In his all too short life of forty-six years he sought to come to some synthesis of the evangelical christianity of his Scottish upbringing and the evolutionary naturalism of the great Victorian expounders of science, Spencer, Huxley, Tyndall and the rest. This interest manifested itself in the nature of the official post he held for a time—lecturer in natural science at the Free Church College in Glasgow. Drummond was as much at home with the aggressive missionaries of the time as with the students of science; thus in 1879 in a visit to America he fitted in a geological tour in the Rockies with Geikie as well as a visit to the evangelical preacher Moody at Cleveland. Later he conducted scientific exploration himself in the lake region of Nyasa and Tanganyika. His writings attracted great attention when they appeared, and some of them became best-sellers, but in his own time they were certainly not understood.

THE NATURALNESS OF THE SPIRITUAL WORLD

When Henry Drummond's most famous book *Natural Law in the Spiritual World* first appeared,[1] in 1883, it had a mixed reception for this reason. The scientific world was on the whole glad to see religious doctrines discussed in scientific terms, or at any rate correlated with scientific ideas, but it was puzzled too, for the treatment of religion was often far from psychological or objective. The religious world was inclined to think that Drummond had in some way or other found support for traditional religious doctrines from the facts revealed by science, but it was very uneasy at finding the "supernatural" treated as a sort of continuation of the "natural." It is safe to say that this famous book has not been much better comprehended at any time since then, though it has been very widely read.

Natural Law in the Spiritual World[2] is a naïve book, but it has the naïveté of something fundamentally true, and something said a long time before people were ready to appreciate it. To-day it must be read with care, for it has many faults. There are far-fetched archaisms, such as the discussion on gravitation[3] or the extension of Spencer's definition of life to "eternal life."[4] There are downright mistakes, such as the too confident discussion of spontaneous generation[5] due to the absence of modern knowledge on the nature of the viruses and similar forms on the borderline of the living and the dead; and again, the persistent attribution of ethical values to the behaviour of the lower animals,[6] a level at which such values are not applicable. Then there are dangerous passages, opening the way to deplorable social activity, perhaps more recognisable in our day than they were in Henry Drummond's. For example, the relations of "Nature" and "Sin"[7] are insufficiently safeguarded against the heresy of those who would build human society purely upon a biological rather than upon a sociological basis. Moreover, we must not be too hasty in thinking that we can recognise either what constitutes sin or what is natural to man. And the remarks on "love of life"[8] err too much in the direction of a grim and joyless puritanism.

But when all criticisms have been made, *Natural Law in the Spiritual World* remains a great book. At first sight, it appears to be devoted to the attempt to show that there are analogies between those natural

[1] Abbreviations adopted in what follows:
    NLSW *Natural Law in the Spiritual World.*
    AOM   *The Ascent of Man.*
[2] Refs. to 47th edition. Hodder & Stoughton, n.d.   [3] NLSW, p. 43.
[4] NLSW, p. 214.   [5] NLSW, p. 64.   [6] NLSW, pp. 319, 344.
[7] NLSW, p. 105.   [8] NLSW, p. 197.

laws discovered by physicists, chemists and biologists, especially the latter, and certain spiritual (or, as we might say to-day "psychological") laws, found to be valid in the realm of religious experience. Drummond was indeed severely criticised for basing his case wholly upon an argument from analogy. But this was a complete misunderstanding of his work. He meant very much more. "The position we have been led to take up," he writes,[1] "is not that spiritual laws are analogous to natural laws, but that they are the *same* laws. It is not a question of analogy, but of identity. The natural laws are not the shadows or images of the spiritual in the same sense as autumn is emblematical of decay, or the falling leaf of death. The natural laws, as the law of continuity might well warn us, do not stop with the visible and then give place to a new set of laws bearing a strong similitude to them. The laws of the invisible are the same laws, projections of the natural, not supernatural."

The implications of this position are indeed far-reaching. Basic to Drummond's ideas was the conception of continuing evolution. This was what he meant by the "law of continuity." The universe consists of a series of levels of complexity and organisation, hierarchical in thought and successive in time, for the simpler preceded the more complicated. First came an evolution of the chemical elements (the "immortal families" of electrons), and the preparation of the stage for the drama of life as the solar systems were formed. In Auden's words:

> "The universe of pure extension where
> Nothing except the universe was lonely,
> For Promise was occluded in its womb
> Where the immortal families had only
> To fall to pieces and accept repair,
> Their nursery, their commonplace, their tomb,
> All acts accessory to their position
> Died when the first plant made its apparition."

Then came the whole long course of biological evolution, leading from the single-celled organism to the primates and man. But still the onward course of organisation did not cease, and man's great and complex brain permitted the development of psychological and above all, of social, organisation. Sexual units united into tribes, tribes into peoples, peoples into nations, and the end of this process is not yet,

[1] NLSW, p. 11.

for the rationally organised world-state of humanity lies still in the future.

> "Through a long adolescence, then, the One
> Slept in the sadness of its disconnected
> Aggressive creatures, as a latent wish
> The local genius of the rose protected,
> Or an unconscious irony within
> The independent structure of the fish;
> But flesh grew weaker, stronger grew the Word,
> Until on earth the Great Exchange occurred."

Now in emphasising all this Drummond was doing nothing so very original. Herbert Spencer and other sociologists had laid great stress on the continuity of biological with social evolution. Karl Marx and Frederick Engels had adumbrated the idea of levels of organisation in setting the Hegelian dialectic actually within evolving nature. But from inside the christian tradition Drummond was working towards these thinkers. His conviction that the "supernatural" of the theologians was, in a sense, supremely "natural," echoed one of the best ideas of the dialectical materialists, namely that materialism had for too long been "misanthropic" or "ascetic" (as Marx had said) and if it were to play its rightful part in future human thought, must do full justice to all the highest aims and strivings of man.

For Marx, materialism, in becoming dialectical, would include all that the christians had meant by the spiritual world. For Drummond, the spiritual world ought to take its place as the highest, but fully natural, level in the evolutionary series. The sublimities of human altruism must thus be thought of, not as something supernatural or mystical, but as characteristic of the highest grades of natural organisation known to us. From this point of view the gulf which so many have imagined to exist between an atheist labour organiser who gives his life for the people, and a christian saint and martyr is so narrowed as almost to disappear. Mediaeval scholasticism had contained some premonitions of Drummond's line of thought. "Gratia non tollit natura," said Thomas Aquinas, "sed supplet et perficit defectum naturae"—Grace does not abrogate nature, but extends and perfects it. Exactly so does the higher level of organisation supersede the lower. And thus we can understand Drummond's phrase "the naturalness of the supernatural."[1] When that is disclosed, and not till then,

[1] NLSW, p. xxii.

"will men see how true it is that to be loyal to all of nature, they must be loyal to that part described as spiritual." If he ever knew of it, he must have approved of the verse ascribed to Thomas Aquinas' contemporary, the Persian Sufi poet Jalal'ud-Din Rumi, who has been called the Thomas à Kempis of Islam:—

> "I died from mineral and plant became
> Died from the plant and took a sentient frame
> Died from the beast and donned a human dress
> When by my dying did I e'er grow less?"

Though referring to the ancient Aristotelian doctrine of the succession of souls (vegetative, sensitive, and rational) in the scale of nature and the development of individual man, and doubtless written in the interest of the doctrine of immortality, these lines do emphasise the human-ness of human beings. Their society cannot be built on sub-human lines. They must be loyal to what Drummond called "the spiritual."

Drummond never tires of describing the levels of organisation, atomic, molecular, crystalline, organic, cellular, organismic, biological, social, etc. "The inorganic had to be worked out before the organic, the natural before the spiritual."[1] In another passage, he correctly shows the relations between the levels of organisation: "It is of course not to be inferred that the scientific method will ever abolish the radical distinctions of the spiritual world. True science proposes to itself no such general levelling in any department. 'Any attempt to merge the distinctive characteristic of a higher science in a lower, of chemical changes in mechanical, of physiological in chemical, above all, of mental changes in physiological, is a neglect of the radical assumption of all science, because it is an attempt to deduce representations of one kind of phenomenon from a conception of another kind which does not contain it, and must have it implicitly and illicitly smuggled in before it can be extracted out of it.' "[1] This is the meaning of the irreducibility of biology to physics and chemistry. It has nothing to do with vitalism in any form; it simply means that the laws which apply at the level of the organic do not operate at the level of the inorganic. But at the same time it must always be remembered that though we can chart out quite fully the laws existing at a given high organisational level, we can never hope to understand how they fit in to the picture of nature as a whole, i.e. how they

[1] NLSW, p. 18.   [2] NLSW, p. 21.

join with the next higher and the next lower levels, unless we are at the same time analysing those levels too. About this there is nothing obscurantist, nothing animistic. Organisation is not inscrutable. Organisation and Energy are the two fundamental problems which all science has to solve.

"It is quite true," writes Drummond,[1] "that when we pass from the inorganic to the organic, we come upon a new set of laws. But the reason why the lower set do not seem to act in the higher sphere is not that they are annihilated but that they are overruled. And the reason why the higher laws are not found operating in the lower is not because they are not continuous downwards, but because there is nothing for them to act upon. It is not law that fails, but opportunity." This overruling of laws characteristic of lower levels, or rather, the application of them in different, more co-ordinated, ways, is a fundamental phenomenon. It constitutes the complete refutation of all fascist philosophers who wish to build human society upon a purely biological basis. Human society must be built upon a sociological basis, and even, as Drummond would have said, upon a spiritual basis.

It is here that we touch upon Drummond's unique contribution to this line of thought, a contribution which perhaps only a christian could have made. Social thinkers such as those who have already been mentioned tended to mould their thought in broad, un-individual terms, dealing with social mechanics and social statistics. But the individual human being is also an entity of a high level of organisation, the psychological. Hence Drummond's insistence on the development of the individual personality. There is a definite affinity between the classless world-state with its ordered production and racial equality, and the *Regnum Dei*, the genuine new world order, the christian ideal of comradeship and social justice. But it was Drummond who pointed out that for this noble state, noble individuals would be necessary, guides and leaders towards it no less than operators of it. He even placed the "natural man" over against the "spiritual man" as a different level of organisation. "The spiritual man,"[2] he wrote, "is removed from the general family of men so utterly by the possession of an additional characteristic that a biologist, fully informed of the whole circumstances, would not hesitate a moment to classify him elsewhere. . . . It is an old-fashioned theology which divides the world in this way, which speaks of men as living or dead, lost or saved, a

[1] NLSW, p. 43.   [2] NLSW, p. 83.

stern theology all but fallen into disuse. The difference between the living and the dead in souls is so unproved by casual observation, so impalpable in itself, so startling as a doctrine, that schools of culture have ridiculed or denied that grim distinction. Nevertheless it must be retained. 'He that hath not the Son hath not Life.' "

At first sight this reads like some voice from the 17th century, and it would indeed have been a mistake to revive any Calvinistic rejection of the weaker brethren. But must we not admit the part which heroic virtue plays in human evolution? Must we not pay homage to that "additional characteristic" which enables some men to help onwards their fellows? For the kind of personality we have in mind, only the Chinese word *Chün-tze* (君子) is adequate, for it means all at once, the learned scholar, the true gentleman, the great-souled man ($\mu\epsilon\gamma\alpha\lambda\acute{o}\psi\upsilon\chi\sigma\varsigma$) of Aristotle, the hero of natural or spiritual war, the magnanimous lover of the people. There are many who spring to the mind in this connection in their various ways—Sir Thomas More, the martyr of mediaeval internationalism; John Lilburne and Thomas Rainborough, fighting intellectuals of the English Revolution; Louis Pasteur, type of the scientific benefactors to humanity; Vladimir Ilyitch Lenin, philosopher and revolutionary leader of the people, visionary of the machine in the service of man; John Cornford, the young historian who thought it worth while to exchange a lifetime of study for the grave of a soldier of the International Brigade in Spain.

It is an interesting commentary on time's changes in thought to read the account of Henry Drummond in the *Dictionary of National Biography*. In 1894 he published some American lectures under the title "The Ascent of Man."[1] On this one of his biographers wrote, somewhat condescendingly, "Drummond's adroitness in rehandling old arguments was truly remarkable, but his general thesis that the struggle for life gradually became altruistic in character, or a 'struggle for the life of others,' and that 'the object of evolution is love' was very severely criticised by men of science, while some of his attempts to qualify the apparent harshness of the scheme of natural selection, by such phrases as 'With exceptions, the fight is a fair fight; as a rule there is no hate in it, but only hunger,' or, 'It is better to be eaten than not to be at all,'' must appear to be perilously near the grotesque." Grotesque, no doubt, to his contemporaries, but as we look to-day at the long aeons of rise in level of biological and social organisation, what else can the trend of evolution be except

[1] (London, 1894).

towards the higher levels of social solidarity which we have not as yet attained? Can we not consider love as the essential social cement, the counterpart at high levels of organisation of the valencies, van der Waals forces and other bonds the operation of which we see among the molecules and colloidal aggregates? Hear Auden again:

> O beggar, bigwig, mugwump, none but have
> Some vision of that holy Centre where
> All time's occasions are refreshed; the lost
> Are met by all the other places there;
> The rival errors recognise their loves
> Fall weeping on each other's necks at last;
> The rich need not confound the Persons, nor
> The Substance be divided by the poor.
>
> Our way remains, our world, our day, our sin;
> We may, as always, by our own consent
> Be cast away: but neither depth nor height
> Nor any other creature can prevent
> Our reasonable and lively motions in
> This modern void where only Love has weight,
> And Fate by Faith is clearly understood,
> And he who works shall find our Fatherhood."

As for the hate and hunger of the struggle for existence in the animal world, the Victorians tormented themselves like so many fakirs with anxiety-neuroses about the tortures of the lower organisms. Into every fish and insect they projected an image of their own highly-organised human minds, endowing the humblest creatures with capacities for fear and pain which a dispassionate consideration of their nervous systems would have shown that they were entirely incapable of. The biologist G. J. Romanes, who, like Henry Drummond, had religious affiliations, though in Romanes' case these were Anglo-Catholic, excelled in perplexities on this score.[1]

There is a real and interesting difference, however, between our attitude to the evolutionary process and that adopted by thinkers such as T. H. Huxley. "Let us understand, once and for all," he wrote,[2] "that the ethical progress of society depends, not on imitating the

[1] G. J. Romanes, *Thoughts on Religion*, ed. by Charles Gore (London, 1895), and *Life and Letters* (London, 1896).
[2] *Evolution and Ethics* (7th edn., London, 1911, p. 83).

cosmic process, still less in running away from it, but in combating it." I shall quote from an acute commentary which my friend C. H. Waddington[1] has written on this pronouncement. "Huxley was writing under the spell of that extraordinary impulsion, so incomprehensible to us to-day, which forced the Victorians to transmute the simple mathematics of their major contribution to theoretical biology into a battleground for their sadism. To Huxley, the cosmic process was summed up in its method; and its method was 'the gladiatorial theory of existence' in which 'the strongest, the most self-assertive, tend to tread down the weaker;' it demanded 'ruthless self-assertion' and 'the thrusting aside or treading down of all competitors.' To us, that method is one which, among animals, turns on the actuarial expectation of female offspring from different female individuals, a concept as unemotional as a definite integral; and we recognise that quite other, though equally natural, methods of evolution may occur when it is societies and not individuals that are in question. Moreover, being no longer hypnotised by the *methods* of evolution we can see its *results*, and these can not be adequately summarised as an increase in bloodiness, fierceness, and self-assertion." The matter could not be better put. Man must indeed combat sub-human nature in so far as it must be subdued to his will. He must never imitate it in forming his social order. Evolution as a whole is neither a scene of Flaubertian tortures as the Victorians saw it, nor yet does it betray any conscious purpose of goodness. But in so far as the highest human societies, yet achieved or yet to be achieved, are the product of evolution, the good has arisen out of the evolutionary process. Just as Drummond said.

His book *The Ascent of Man* is of such interest that we must discuss it a little further. He did not merely draw attention to the earliest phases of co-operation exhibited in the coming together of cells to form the metazoan organisms, or in the development of close animal associations like the termites and ants. He successfully traced the origins of social altruism and co-operativeness to the many-sided phenomena of parental care, and ultimately to that donation of part of the self for the benefit of the offspring which occurs in every reproductive act.[2] While recognising the principle of natural selection and the struggle for existence as the main factor in the origin of

[1] Nature, 1941.
[2] AOM, 1902 edition, pp. 278, 282, 286, 398. In many cases this takes extreme forms, e.g. the exhaustion of the muscles of the migrating salmon in egg formation.

species, he insisted that it was only half the story.[1] He was bold enough to conclude, therefore, that love was not a late arrival, an afterthought, in evolution, but in a sense the goal of the whole process since by its operation alone could the highest stages of social human organisation come into being.[2] "There is no such thing in nature," he remarked, "as *a* man." There is only social man.[3] Furthermore, Drummond recognised in contemporary human society certain anti-social forces, which he designated as "War" and "Industry" (by which he implied predatory competition); these he believed were the remains of the struggle for existence at sub-human levels. What he says on this subject is so interesting that it deserves quotation in full:—

> "When we pass from the animal and the savage states to watch the working of the struggle for life in later times, the impression deepens that, after all, the 'gladiatorial theory' of existence has much to say for itself. To trace its progress further is denied us for the present, but observe before we close what it connotes in modern life. Its lineal descendants are two in number, and they have but to be named to show the enormous place this factor has been given to play in the world's destiny. The first is *War*, the second is *Industry*. These in all their forms and ramifications are simply the primitive struggle continued on the social and political plane. . . . Along with Industry and for a time before it, War was the foster-mother of civilisation. . . . When society wonders at its labour troubles it forgets that Industry is a stage but one or two removes from the purely animal struggle.
>
> But one has only to look at the further phases of the struggle to observe the most important fact of all, the change that passes over the principle as time goes on. Examine it on the higher levels as carefully as we have examined it on the lower, and though the crueller elements persist with fatal and appalling vigour, there are whole regions, and daily enlarging regions, where every animal feature is discredited, discouraged, or driven away. . . .
>
> The amelioration of the struggle for life is the most certain prophecy of science. . . . We find the animal side of the struggle for life attacked in such directions, and with such weapons that its defeat is sure. These weapons are in the armoury of nature;

[1] AOM, pp. 30 ff.   [2] AOM, pp. 276 and 428.   [3] AOM, p. 312.

they have been there from the beginning; and now they are engaged upon the enemy so hotly and so openly that we can discover what some of them are. The first is one which has begun to mine the struggle for life at its roots. Essentially, as we have seen, this struggle is the attempt to solve the fundamental problem of all life—Nutrition. If that could be solved apart from the struggle for life, its occupation would be gone. That problem will be solved by science. At the present moment chemistry is devoting itself to the experiment of *manufacturing nutrition*, and with an enthusiasm which only immediate hope begets.... 'The time is not far distant when the artificial preparation of articles of food will be accomplished....'

But there is a higher hope. As there comes a time in a child's life when coercion gives place to free and conscious choice, the day comes to the world when the aspirations of the spirit begin to compete with, to neutralise, and to supplant, the compulsions of the body. Against that day, in the heart of humanity, nature had made full provision. For there, prepared by a profounder chemistry than that which was to relieve the strain on the physical side, had gathered through the ages a force in whose presence the energies of the animal struggle are as nought. Beside the struggle for the life of others, the struggle for life is but a passing phase. As old, as deeply sunk in nature, this further force was destined from the first to replace the struggle for life, and to build a nobler superstructure on the foundations which it laid."[1]

The passage is interesting for many reasons, of which two may be mentioned. First, Drummond put his finger on the very power which has changed and will change the face of human civilisation, applied science. His thought was exactly parallel with that of Marx and Engels, who some years before had exchanged letters describing their reading of the books of Liebig and the other agricultural chemists, and showing how the rise of nutritional productivity which science was bringing about knocked Malthus into a cocked hat. But secondly though Drummond could recognise reproductive donation and parental care as the origin of social altruism, he could not pass over the threshold crossed by Marx and Engels alone,[2] and point with absolute clarity to the working-class movement of his day as the historic force destined to lead the way to the higher forms of social

[1] AOM, pp. 269 ff.   [2] AOM, cf. p. 431.

organisation, and to bring humanity from the Egypt of necessity into the Canaan of freedom.

It was mentioned above that love might be considered as the analogue of the physical bonds which unite particles at the molecular level. This was indeed an idea of Henry Drummond's. Though it would have been acceptable to the ancient Ionian scientists, it was doubtless grotesque enough to the Victorians.

> "Is it conceivable," he wrote, "that in inorganic nature, among the very material bases of the world, there should be anything to remind us of the coming of the Tree of Life? To expect even foreshadowings of ethical characters there were an anachronism too great for expression. Yet there is something there at least worth recalling in the present connection.
>
> The earliest condition in which science allows us to picture this globe is that of a fiery mass of nebulous matter. At the second stage it consists of countless myriads of similar atoms, roughly outlined into a ragged cloud-ball, glowing with heat, and rotating in space with inconceivable velocity. By what means can this mass be broken up, or broken down, or made into a solid world? By two things, mutual attraction and chemical affinity.... What affinity even the grossest, what likeness even the most remote, could one have expected to trace between the gradual aggregation of units or matter in the condensation of a weltering star, and the slow segregation of men in the organisation of societies and nations? However different the agents, is there no suggestion that they are different stages of a uniform process, different epochs of one great historical enterprise, different results of a single evolutionary law?"[1]

Thus as cosmic development proceeded, conditions arose in which highly complex molecular aggregates became possible. Various forces joined them together. All things have come into being by way of an eternal battle between attraction and repulsion, aggregation and disaggregation, in which the victories of aggregation, though decisive, are never absolutely complete; remnants of the defeated remaining as essential elements of the new level of organisation. Thus are the Furies conducted to their cave under the Acropolis; and the Dragons incorporated into the Civil Service.

We are reminded of the Orphic hymn to love in Longus' *Daphnis*

[1] AOM, p. 432.

and Chloe: "My young friends," says Philetas, "Love is a god, young, beautiful, and ever on the wing. He therefore rejoices in the company of youth, he is ever in search of beauty, and gives wings to the souls of his favourites. His power far exceeds that of Zeus. He commands the elements; he rules the stars; he governs the world, and even the gods themselves are more obedient to him than these your flocks are to you. All these flowers are the works of Love, these plants and shrubs are his offspring; through him these rivers flow and these zephyrs breathe."[1] But all this is of course poetry. We cannot in sober truth be such hylozoists. One must remember that historically the attribution of animistic emotions, etc., to inorganic things was the primitive method of explanation of natural processes, from which science freed itself in the 17th and 18th centuries only with great difficulty.[2] Still, we can say with Drummond that there may be something analogous between the bonds appropriate to each of the different levels of organisation in the world. And we remember that greak book in which Sigmund Freud described what he called "the task of Eros" in the process of human social evolution.[3] And hence the true profundity in the repeated invocations (so beautiful, but sometimes in the past to me so puzzling) of Lucretius to the Epicurean Venus, *alma Venus*:—

"quae quoniam rerum naturam sola gubernas
nec sine te quicquam dias in luminis oras
exoritur neque fit laetum neque amabile quicquam"[4]
(. . . For thou alone
Governest all things, and without thee naught
Is risen to reach the shining shores of light,
Nor aught of joyful or of lovely born.)

She who is the Delight of Gods and Men embodies the principle of Union and Aggregation.

Thus it will be seen that as social evolution is continuous with biological evolution, so (Drummond was convinced) much of the content of traditional christian theology—"the laws of the spiritual world"—arose directly from what had preceded it in the highly organised realm of the psychological. In this he complemented in an important way the thought of his contemporaries, though his

[1] Geo. Thornely's translation, 1657.
[2] Cf. J. G. Gregory "The Animate and Mechanical Models of Reality," Journ. Philos. Stud., 1927, **2**, 301.
[3] *Civilisation and its Discontents* (London, 1930).   [4] *De Rer. Nat.* I, 21.

naturalism was so revolutionary that in his own time it was not well understood. Moreover, he fully appreciated the fundamental point that evolution is not over, and that our present social system is not the best of which man, and hence nature, is capable. "Why should evolution," said Drummond,[1] "stop with the organic?" "These kingdoms," he concludes his book, "rising tier above tier in ever increasing sublimity and beauty, their foundations visibly fixed in the past, their progress, and the direction of their progress, being facts in nature still, are the signs which, since the Magi first saw his star in the east, have never been wanting from the firmament of truth, and which in every age with growing clearness to the wise, and with ever-gathering mystery to the uninitiated, proclaim that the Kingdom of God is at hand." This is surely only another way of saying that the new world order of social justice and comradeship, the rational and classless world state, is no wild idealistic dream, but a logical extrapolation from the whole course of evolution, having no less authority than that behind it, and therefore of all faiths the most rational.

It costs something to say this, for these words are written after the news of the outbreak of what must come to be called the second world war. It is incredible that the agonies of 1914–1918 are again to be repeated. But even so gigantic a set-back as a war of this magnitude unleashed by fascism cannot shake a faith which is based on the considerations which convinced Drummond and Spencer, Engels and Marx. The way may be long and we may not live to see, but the triumph of the rational spiritual man is sure.

S.S. President Harding [*Lone Star State*]
At sea, September 6, 1939

[1] NLSW, p. 401.

Science, Religion and Socialism

(A contribution to the book of essays *Christianity and the Social Revolution*, 1935; with additions, including material from the Criterion, 1932, and Scrutiny, 1932.)

THE problem of the relationship between the traditional religion of the European West and the coming new world-order, as yet in its details uncertain, seems at first sight to have little to do with the preoccupations of the scientist. Whether the old forms of theology and liturgy disappear, whether the new social order is, or is not, more just than that which is breaking up, whether he has to live and work in the corporate or in the classless State, might seem to be matters of indifference to him. Nevertheless such a view would be superficial. The moment a scientific worker begins to reflect upon the nature and methods of his science, he will find himself involved in its history and philosophy, and hence its relations to historical, economic and intellectual factors, from which religious ideas certainly cannot be excluded. The moment he begins to reflect upon the ends to which others are putting the results of his work, he will find himself involved in the current political discussion of his time. Even some hypothetical scientist who aimed at the most complete neutrality with respect to the world in which he lived could not long escape the ultimate argument of economic forces, and would be induced to think over his relation to his fellows when he found himself unemployed after some sudden restriction of scientific effort.

The beginnings of the scientific movement in the 17th century are discussed elsewhere in this book. Acquisition of personal wealth, the fundamental motive of capitalist enterprise, acted then, and for a long time afterwards, as the most powerful stimulus and support for scientific research. But the indiscriminate application of the scientific method to natural things bursts in the end these limitations. It shows us not only how to make textiles and cheese, but also how, if we will, a high degree of universal physical and mental well-being may be achieved. In so doing, it goes beyond the facts which any single group of men can lay hold of with the object of acquiring private riches. And it dictates to the scientific worker a new allegiance, a separation from his allies (or masters) of three centuries' standing.

The Position of the Scientific Worker.

The position of the scientific worker in the world of to-day is indeed a very difficult one. Owing to the gradual permeation of our entire civilisation by the practical results of scientific thought and invention, the scientific worker has in some measure succeeded to the semi-oracular tripod previously occupied by the religious thinker, whether enthusiastic saint or prudent ecclesiastic. That ancient separation of life into secular and sacred, which arose out of the acquiescence of the early christians in their failure to transform the human society of their time into God's Kingdom on earth, still reigns in our civilisation. Owing to the increasing intellectual difficulties which the ordinary man of our time feels with respect to the theology of the traditional form of western European religion, he turns more and more to the scientific worker, expecting to hear from *him* a sound doctrine about the beginning of the world, the duty of man, and the four last things. The scientific "ascetic" in the laboratory is the monk of to-day, and is tacitly regarded as such by the ordinary man.[1] Conversely, the secular power, the medieval *imperium*, has been succeeded by the power of the owner—the owner of factories, the owner of newspapers and propaganda agencies, the owner of land, the owner of finance capital.

In a new guise, then, the sacred and the secular are still at war. We may study their antagonism best by observing the fate of the concept of *Regnum Dei*, the Kingdom of God—always the surest indication of the relative power of priest and king. Roughly speaking, there have been, in the history of the Christian Church, three separate doctrines about the Kingdom of God, three separate interpretations of the Kingdom-passages in the Gospels.[2] First, there was the identification of the Kingdom with a purely spiritual mystical realm of beatitude, either to be reached after death by the faithful, or attainable here and now through the methods of prayer and ascetic technique, or existing in the future in Heaven after the last judgment. This has been perhaps the commonest theory. It has flourished whenever the secular was strong, since it discountenanced any attempt to improve the conditions of life on earth. As an instance, one could mention the mystical theology of lutheranism, whose founder held the world

[1] He may be called an "ascetic" in that he has often sacrificed for his intellectual calling those material benefits which Lord Birkenhead referred to as the "glittering prizes" of the capitalist system.

[2] Cf. Bishop A. Robertson, *Regnum Dei* (London, 1901). We shall discuss this subject in more detail below, p. 50.

to be utterly bad and irredeemable, a realm of Satan, from which the only escape was by means of religious exercises within the organised body of christians.[1] But, secondly, in every age there have been those who have interpreted the Kingdom as a state of divine justice in the future and in the world, to be attained by unceasing effort on the part of men and women. This struggle was the outcome of their thirst for social justice, and gave meaning to all martyrdoms since the beginning of the world.

In ages when ecclesiastical organisation was powerful, the visible Church itself, sharing the world with the temporal emperor in a condominium, could be identified with the Kingdom of God. This was a third interpretation. With the reformation and the splitting of the universal church into a thousand sects it lost its force.

But if the scientific worker is the modern representative of the mediaeval cleric, he finds himself in a relatively much worse position. Science in our time is not able to dictate its terms to capitalist "captains of industry" and the governing class in general; on the contrary, it is in utter bondage, dependent upon their fitful and grudging support, itself divided by dangerous national boundaries and sovereignties. In such a case we should expect that many scientists would interpret the concept of the Kingdom (though none of them, of course, would dream of referring to it under that name) as something spiritual, something harmless, something incapable of any affront to a capitalist world.

This is exactly what we find. Nothing could better illustrate the point than the Huxley Memorial Lecture of A. V. Hill, in 1933, and his subsequent controversy with J. B. S. Haldane—two of England's most distinguished biologists.[2] The discoveries of science, said Hill, whatever mistakes may be made, do gradually build up a structure which is approved by all sane men; in the last three hundred years, the experimental method, which is universal, has produced results beyond all previous human achievements. This universality of its method and results gives science a unique place among the interests of mankind. But "if scientific people are to be accorded the privileges of immunity and tolerance by civilised societies, they must observe

[1] *See* Pascal, R., *The Social Basis of the German Reformation: Martin Luther and His Times* (Watts, London, 1933).
[2] Hill, A. V., Huxley Memorial Lecture, 1933; abridged version: "International Status and Obligations of Science," Nature, 1933, **132**, 952. Hill, A. V., and Haldane, J. B. S., Nature, 1934, **133**, 65.

the rules." "Not meddling with morals or politics; such, I would urge," he went on, "is the normal condition of tolerance and immunity for scientific pursuits in a civilised state." Nothing would be worse than that science should become involved with emotion, propaganda, or particular social and economic theories. In other words—"My kingdom is not of this world," must be taken as meaning not *in* this world either. Let unemployment, repression, class justice, national and imperial wars, poverty in the midst of plenty, etc., etc., continue and increase; nothing is relevant to the scientific worker, provided only his immunity is granted—immunity to pursue his abstract investigations in peace and quiet. Here we substitute for the kingdom-concept of mysticism a kingdom-concept of mathematics, equally sterile with respect to human welfare, equally satisfactory to the powers of this world.

"The best intellects and characters, not the worst," continued Hill, "are wanted for the moral teachers and political governors of mankind, but science should remain aloof and detached, not from any sense of superiority, not from any indifference to the common welfare, but as a condition of complete intellectual honesty." Haldane was not slow to point out that Hill's sterilisation of the scientific worker as a social unit arose from the facile ascription to him of no loyalties save those of his work. In so far as he is a citizen as well as a scientist, he *must* meddle with morals and politics. But Hill's point of view can be attacked more severely from a deeper standpoint. Science does not exist in a vacuum; scientific discoveries are not made by an inexplicable succession of demiurges sent to us by Heaven; science is, *de facto*, involved with "particular social and economic theories," since it exists and has grown up in a particular social and economic structure. Here there is no space even to outline the marks which theoretical and applied science bears revealing its historical position. I merely wish to point out that it is not altogether surprising that the ordinary man expects some lead from the scientific worker in his capacity of citizen. In the Middle Ages, life was ruled by theology, hence the socio-political influence of the theologian; today it is ruled by science, hence the socio-political importance of the scientific worker.

The Treason of the Scholars.

Hill's conception of the Kingdom as a realm of truth and exact knowledge far removed from the affairs of human life has been most clearly formulated in our time by Julien Benda, in his book, *La Trahison des Clercs*.[1] The betrayal of our generation by the clerks, that is to say, by the scientists and scholars which it has produced, he conceives to consist in the fact that whereas the mediaeval clerk was wholly devoted to the working out of the implications of a transcendent truth, the modern clerk has no similar task, and therefore engages without hesitation in the political struggles of the time. "Our century," says Benda, "will be called the century of the intellectual organisation of political hatreds. That will be one of its great claims to fame in the history of human ethics." But does not Benda misread the attitude of the mediaeval clerk? Preoccupied by transcendent truths he might certainly be, but he was also very much concerned about economic relationships, and, by virtue of that fact alone, he *was* politically minded in the modern sense of the words. For modern politics bear no relation to the politics of the mediaeval world. A 13th-century theologian might well leave on one side the quarrels of petty princes about territorial jurisdiction or feudal honours, but he, on his own assumptions, could not, and did not, leave on one side the detailed economics of the commerce and finance of the time. Benda fails to realise that in our days there is no longer any distinction between politics and economics. What are the ferocious modern nationalisms which he describes with such force but devices engineered and operated by economic interests which do not wish co-operation and friendship between the common peoples of the world? What is jingo imperialist patriotism but an instrument designed to drown the call to union of the *Communist Manifesto?*

The mediaeval scene was supremely characterised by its subordination of other interests to religion. We may call it a period of religious genius, when all poetry, literature, learning, and music was co-opted into the service of this primary preoccupation of men. And since this was the case, no human interests could be regarded as outside the sphere of theology, least of all the interests of the market-place, where every economic transaction was a possible opportunity for the snares of the devil, or, alternatively, could, by right arrangement, be turned into an exercise of spiritual profit. The life of man on earth

---

[1] Benda, J., *La Trahison des Clercs* (Grasset, Paris, 1927). The word clerk meant originally any man who could read, an attainment chiefly confined in the Middle Ages to ecclesiastics major and minor, cf. the Book of Common Prayer: "the priest and clerks."

was regarded not as an end in itself, but as the preparation for a fuller life in Heaven, a fuller life which could not be entered into without the passport of justice, temperance, and piety. It was the province of theology, therefore, to regulate public economic affairs just as much as those of individual devotion. The most important means by which this was done were first, the principle of the just price, and, secondly, the prohibition of usury. Every commodity had its just price, based on the cost of its production, and allowing to its producer a margin of profit sufficient for him to live in that degree of comfort which was considered appropriate to his station. It was unchristian to force prices up in a time of scarcity, and thus to take advantage of the necessity of others; unchristian to allow prices to fall in time of glut, and so defraud honest merchants. Usury was prohibited alike by civil and canon law.[1] And the names of many other long obsolete misdemeanours, such as regrating, forestalling and engrossing, remain to show how the theologian systematised mediaeval economic transactions.[2]

What would happen to our present social structure, we might ask, if by some miracle the mediaeval Church were to have full power again, and all usury were prohibited, the principle of the just price exacted, and the restriction of profits renewed? We should, of course, observe a very spectacular collapse. The Middle Ages had, in fact, their own conception of collectivism, but it was fundamentally non-equalitarian. Each group, ecclesiastical, military, or commercial, held a distinct place in a system of social orders possessing different degrees of wealth and social prestige. And although it was true that each order had definite duties towards the other orders, not excluding even the peasant basis, it was equally true that these obligations were frequently unfulfilled. Still, mediaeval society was organic, rather than individualistic and atomic. As Chaucer's Parson said:

> "I wot well there is degree above degree, as reason is, and skill it is that men do their devoir thereas it is due, but certes extortions and despite of your underlings is damnable."

[1] Cf. W. Cunningham's *Christian Opinion on Usury* (Edinburgh, 1884).

[2] Regrating was the practice of buying goods in order to sell them again in the same market at a higher price, and without adding to their value. Forestalling was the purchase of goods on their way to the market, or immediately on their arrival, or before the market had properly opened, in order to get them more cheaply. Engrossing was the mediaeval counterpart of cornering, the buying up of the whole, or a large part, of the stock of a commodity in order to force up the price.

"The clerk can only be strong," says Benda, "if he is fully conscious of his nature and his function, and if he shows us that he is conscious of it, that is to say, if he declares to us that his kingdom is not of this world. This absence of practical value is precisely what gives greatness to his teaching. As for the prosperity of the kingdoms of this world, that belongs to the ethic of Caesar, and not to his ethic. And with this position the clerk is crucified, but he has won the respect of men and they are haunted by his words." Yet if one of Julien Benda's mediaeval clerks were placed in our modern world, would he not denounce the fantastic system of economics under which we live; would he not criticise the laws which cause food to be destroyed because people are too poor to buy it? It is well that Benda castigates the modern clerk for lending the weapons of his intellect to nationalism, but there are other forces than nationalism at work in the political world to-day. He can, of course, remain inactive, adopting the position of absolute neutrality laid down by Benda, and urged, as we have seen, by distinguished representatives of science, refusing to take part in the political and social struggle, and finally perishing, like an Archimedes, at his laboratory bench during a war. But what differentiates the position of the modern from the mediaeval clerk is that, if he was to be active, the latter had no choice in his allegiance, while the former has a choice, and must make it. Thus there are two ways open to the scientific worker at the present time. All the backward pull of respectability and tradition urges him to throw in his lot with the existing capitalist order, with its corollaries of nationalism, imperialism, militarism, and, ultimately, fascism. On the other hand, he can adopt the ideals of social justice and of the classless State; he can recognise that his own best interests lie with the triumph of the working-class, the only class pledged to abolish classes; in a word, he can think of the Kingdom *literally* and can work for its realisation. A Kingdom not of this world, but to be in this world.

The transcendent truth of the mediaeval Church was bound up with a definite economic order, feudalism; and it was capitalism, of course, as it gradually developed, which upset this economic order, and science which superseded this transcendent truth. The geographical discoveries, which made the European home begin to seem a prison; the astronomical discoveries, which made the earth as a whole, previously the scene of the drama of redemption, shrink to one among a vast number of celestial bodies; the mechanical discoveries which opened up the possibility of industrialism; all undermined the strength of the old-fashioned system until hardly anything was left of it. Moreover, there were discoveries in the spiritual world, too; there was the important protestant discovery that material riches, far from being a presumptive sign of ill dealings, were an outward and visible sign of the inward approval and blessing of God. And most interesting of all, there was the rise of the concept of scientific law, often conceived of in a crude mechanical way, as was only natural at its first beginnings. Who would connect with this the decline in the cult of the Blessed Virgin? Yet there was a certain connection. "The Virgin," wrote Henry Adams,[1] "embarrassed the Trinity. Perhaps this was the reason why men loved and adored her with a passion such as no other deity has ever inspired. Mary concentrated in herself the whole rebellion of man against fate; the whole protest against divine law; the whole contempt of man for human law as its outcome; the whole unutterable fury of human nature beating itself against the walls of its prison-house, and suddenly seized by a hope that in the Virgin there was a door of escape. She was above law; she took a feminine pleasure in turning hell into an ornament, as witness the west window at Chartres; she delighted in trampling on every social distinction in this world and in the next. She knew that the universe was as unintelligible to her, on any theory of morals, as it was to her worshippers, and she felt, like them, no sure conviction that it was any more intelligible to the creator of it. To her, every suppliant was a universe in himself, to be judged apart, on his own merits, by his love for her—by no means on his orthodoxy or his conventional standing in the Church, or his correctness in defining the nature of the Trinity." What a collapse it was when men came to feel that this way of escape was no longer open to them. As canon law decayed, as confidence in the absoluteness and divine authority of civil law disappeared, so scientific law arose like the growing light of day. The mediaeval worship of Mary, so charming, so naïve, was a phenomenon of childhood. She could perhaps save a suppliant from a ruling, a decretal, or a codex, but not from the laws of gravitation or thermodynamics. Mankind was now to take up again the guidance of old Epicurus—

> "Hunc igitur terrorem animi tenebrasque necessest
> non radii solis, neque lucida tela diei
> discutiant, sed naturae species ratioque."[2]

[1] Henry Adams, *Mont St. Michel and Chartres*, Massachusetts Historical Society, p. 276.
[2] *De Rer. Nat.* VI, 39

> (These terrors then, this darkness of the mind,
> Not sunrise with its flaring spokes of light,
> Nor glittering arrows of morning can disperse,
> But only Nature's aspect and her law.)

The path lay open now towards a surer freedom, if first necessity could be understood. Amid such vast changes of intellectual climate, it is not surprising that the function of the clerk should both change and yet remain the same.

The Concept of the Kingdom.

The concept of the Kingdom is of such importance for every aspect of the relations between christianity and communism that I must amplify a little what I said above about the forms which it has taken in christian thought. We may divide the logical possibilities into four. The Kingdom of God has been thought to exist:

(1) Here and now;
(2) Here but not yet;
(3) Not here but now already;
(4) Not here and not yet.

Clearly the most fundamental distinction lies between those who have looked for the Kingdom on earth, whether now or in the future, and those who have interpreted it as meaning an essentially invisible and other-worldly state. The extremest division lies between the second and third alternatives.

The early Church, which for this purpose must be taken as meaning up to the end of the 3rd century in the east and the end of the 4th in the west, was almost wholly devoted to the second of these interpretations. It was believed that the second coming of the Lord, which was thought to be imminent, would inaugurate a visible reign of complete righteousness, in which the saints would administer, until the last judgment, a society based on love and justice. This doctrine, known to theologians as millenniarism, chiliasm, or "realistic eschatology," found its canonical authority largely in the Apocalypse of John, and its intellectual defenders in such men as Cyprian, Justin, Irenaeus and Tertullian. It was attacked, as time went on, by three principal factors. First, there was the necessity of adapting the prophetic vision of a world made new to a world in which the expected leader did not return. Secondly, there was the influence of Hellenistic mysticism and allegorisation, which in the hands of Origen and other more thorough-going neo-platonists, tended to emphasise the third interpretation, i.e. that the Kingdom was a purely mystical idea, existing now but elsewhere, wholly in the world of the spirit. Thirdly, there was the increasing organisation of the Church, and the acceptance of this by the secular power in the time of Constantine; this invited men to diminish their ideals of love and justice, and to identify the Kingdom with an actually existing society. This led to the first interpretation. The Kingdom was "here and now," either in the form of the Eastern Empire or the Latin Church, which after Augustine claimed, and still claims officially to this day, to be itself the Kingdom. Lastly in all the ages of christianity there have been supporters of the fourth and most utterly remote interpretation, namely that the Kingdom means only the reign of God after the last judgment.

The millenniarist viewpoint was essentially a continuation of the great strain of Hebrew prophecy, with which all the actors in the drama of the Gospels, whether known or unknown to us, had certainly been familiar. In this the reality of the time process was quite central. Take, for example, the following passage from Amos:[1]

"Hear this, O ye that would swallow up the needy and cause the poor of the land to fail, saying, When will the New Moon be gone that we can sell corn? and the Sabbath, that we may set forth wheat? making the measure small and the payment great, and dealing falsely with balances of deceit; that we may buy the poor for silver and the needy for a pair of shoes. . . . The Lord hath sworn by the excellency of Jacob, Surely I will never forget any of their works. . . . I will slay the last of them with the sword; there shall not one of them flee away. Though they dig into hell, thence shall mine hand take them; and though they climb up into heaven, thence will I bring them down. . . . But in that day I will raise up the tabernacle of David that is fallen, and close up the breaches thereof; and I will raise up his ruins and build it as in the days of old. Behold, the days come, saith the Lord, that the ploughman shall overtake the reaper, and the treader of grapes him that soweth seed; and the mountains shall drop sweet wine, and all the hills shall melt." It is extremely interesting to contrast the Hebrew apocalyptic conviction that in the future evil will be overthrown and the earth become a common and bountiful treasury for a right-loving people, with the characteristically Hellenic belief in a former Golden Age from which humanity has for ever

[1] Chapter viii. 9.

fallen away. The only other ancient literature which has resemblances to that of the Hebrews in this respect is the Chinese, where remarkable descriptions of social evolution occur in the *I-Ching*, the "Book of Changes" (4th or 5th century B.C.). Hesiod, on the other hand, says that if it had not been for the act of Prometheus, who stormed heaven by force, brought thence the gift of fire, and provoked the gods to withhold from men an easy way of life, "you would have been able to do easily in a day enough work to keep you for a year, to hang up your rudder in the chimney corner, and let your fields run to waste."

Thus have decayed the first bright hopes and visions of the christians. In a most interesting passage, Bishop Robertson reveals the class character of the opposition to millenniarism. "Intense as was the christian instinct to which millenniarism gave articulate form, it was in some respects in latent antipathy to the ecclesiastical spirit, and waned as that spirit gathered strength. Its rejection by rational theology, and by the trained theologians who filled the more important places in the Greek Churches in the third and fourth centuries, had practically the effect of ranging the clergy in opposition to it. In fact, millenniarism, by virtue of its direct appeal to minds of crass simplicity, was a creed for the lay-folk and the simpler sort. When religious interest was concentrated upon it, it would indirectly undermine the interest felt in doctrines requiring a skilled class to interpret them. The apocalyptic spirit is in fact closely akin to the spirit of unregulated prophesying, and the alliance has been apparent, not only in the second century, but in mediaeval and modern times as well." Crass simplicity—might we not almost say inferior economic position? A skilled class—perhaps a privileged one too?

Of the hopes of the "simpler sort" we get a glimpse in that very interesting fragment of Papias, preserved by Irenaeus[1] and believed to be an authentic saying of Christ himself,[2] "The days will come when vines shall grow, each having ten thousand branches, and on each branch ten thousand twigs, and on each twig ten thousand shoots, and on each one of the shoots ten thousand clusters, and on every cluster ten thousand grapes, and every grape when pressed will give twenty-five firkins of wine. And when any one of the saints shall lay hold upon a cluster, another shall cry out, 'I am a better cluster; take me; bless the Lord through me.' And in like manner, that a

[1] See the *Apocryphal New Testament*, ed. M. R. James (Oxford, 1924, p. 36).
[2] "Old men who knew John the Lord's disciple, remember that they heard from him how the Lord taught concerning those times, saying, etc."

grain of wheat will produce ten thousand stalks, and each stalk ten thousand ears, ..." and so forth.[1] It has often been said that the communism of the early christians was purely one of distribution, not of production. Here, however, we have, as it were, a dream of the abundance of natural wealth latent in the world's productive forces, and to be unloosed by science so many centuries later.[2] But the inevitable answering note is struck. Asceticism comes to the aid of the possessing classes, and when we turn to Augustine we find: "The opinion that the saints are to rise again would at least be tolerable if it were understood that they would enjoy spiritual delights from the presence of the Lord. We ourselves were formerly of this opinion. But when they say that those who then arise will spend their time in immoderate carnal feastings—in which the quantity of food and drink exceeds the bounds not only of all moderation, but of all credibility—such things cannot possibly be believed except by carnal persons."

Whatever happened in later centuries, then, it is certain that the christians of the primitive church put their Kingdom on the earth and in the future. To this belief of "crass simplicity" let us return. We reach the paradox that Marx and Engels would have been more acceptable to the martyrs and the Fathers than the comfortable 19th century theologians contemporary with them, seeking to excuse and support the phenomena of class oppression. For the kingdom of Marx was not of this world, but to be in this world.

Yet Benda goes on: "I regard as being able to say 'my kingdom is not of this world' all those whose activities do not pursue practical ends, the artist, the metaphysician, the scientist in so far as he finds satisfaction in the exercise of his science and not in its results. Many will tell me that these are the true clerks, much more than the christian, who only embraces the ideas of justice and love in order to win salvation." Here he adopts, as I think, a quite unjustifiable separation of these activities from practical affairs. In science, at any rate, the closest relations exist between practical technology and pure research.

---

[1] Similar accounts occur in the Jewish *Apocalypse of Baruch* and the Coptic *Apocalypse of James*.

[2] "So when the Lord was telling the disciples about the future kingdom of the saints, how glorious and wonderful it would be, Judas was struck by his words and said, 'Who shall see these things?' And the Lord said: 'These things shall they see who are worthy.'" (Hippolytus, *On Daniel*, 4.) "Papias says that when Judas the traitor believed not and asked, 'How then shall these growths be accomplished by the Lord?', the Lord said: 'They shall see who shall come thereto.'" (Irenaeus, *Contra Haer*, 5.)

Biology would be in an etiolated condition if it were not bound up at every point with stockbreeding, agriculture, medicine, the fisheries, and sociology. With physics and chemistry the case is even more obvious. "Historically, the sciences grow out of practice, the production of ideas arises out of the production of things."[1] It is true that in science we must not set out, in general, to solve problems *because* the answer will afford some new invention, but it is often the technical practice which suggests the problem. The great difference which we must recognise between mediaeval theology and modern science, is that an economic structure was directly and logically derivable from the former, and no clear system in such matters has as yet arisen from the latter. The former incorporated a system of ethics, in the form of moral theology. The latter has not as yet produced one.

Where, then, is the moral theology of to-day? The only possible answer is that communism provides the moral theology appropriate for our time.[2] The fact that a doctrine of God is apparently absent from it is unimportant in this connection; what it does is to lay down the ideal rules for the relations between man and man, to affirm that the exploitation of one class by another is immoral, that national wars for markets are immoral, that the oppression of subject and colonial races is immoral, that the unequal distribution of goods, education, and leisure is immoral, that the private ownership of the means of production is immoral. It dares to take the "love of our neighbour" literally; to ensure that by the abolition of privilege each single citizen shall have the fullest opportunities to live the good life in a community of free and equal colleagues. It continues and extends the historic work of christianity for woman, setting her on a complete equality with man. Its concept of leadership is leadership from within, not from above.

Only because christian theology three centuries ago gave up the attempt to apply a very similar ethic to human affairs has this state of things come about. The essential weakness of the modern clerk resides in the fact that vast progress in art or science appears at first sight to be theoretically equally compatible with national capitalism or with international communism. The economic doctrines which he

[1] Bukharin, N., *Theory and Practice from the Standpoint of Dialectical Materialism* (Kniga, London, 1931, p. 5).

[2] Cf. for example the essay "Communism and Morality" by A. L. Morton in *Christianity and the Social Revolution*, 1935, and "Marxism and Morality" by J. Hunter in University Forward, 1941, 6, 4.

must adopt are not at first sight a direct consequence of his own fundamental axioms, but embody themselves in a social theory external to his own sphere. Hence the dual character of the scientific worker, as scientist and as citizen. Hence the temptation for him to shirk his public responsibilities and as "pure clerk" to be silent except when he gives the results of his own exact researches.

We may remember the bitter words said to have been prefixed by the mathematician, G. H. Hardy, to a book on pure mathematics: "This subject has no practical value, that is to say, it cannot be used to accentuate the present inequalities in the distribution of wealth or to promote directly the destruction of human life."

Perhaps the most important task before scientific thinkers to-day is to show in detail how the ethics of collectivism do in fact emerge from what we know of the world and the evolutionary process that has taken place in it. Scientific socialism (I believe) is the only form of socialism which has the future before it; its theoreticians must therefore show not only that high levels of human social organisation have arisen and will arise by a continuation of the natural process, but what are the ethics appropriate to them. Scientific ethics should be to communist society what catholic ethics were to feudal society and protestant ethics to capitalist society.

Theology and the Modern Man.

In the preceding section I said that a doctrine of God was apparently absent from communist thought. I used the word "apparently" because (a) dialectical materialism might be logically compatible with a spinozistic theology;[1] (b) the immanence of the christian Godhead as Love is better provided for in communism than in any other order of human relationships. Future communist Clements of Alexandria will have the task of codifying the *praeparatio evangelica* of the christian centuries.

Today we are all Taoists and Epicureans. For the taoists, the Way of Nature was *tzu-jan* (自 然); it came *of itself*. So also in Lucretius' great poem[2]:

". . . natura videtur
libera continuo dominis privata superbis
ipsa sua per se sponte omnia dis agere expers."

---

[1] Cf. *Moscow Dialogues* by J. Hecker (London, 1933), p. 55, and *Fundamental Problems of Marxism*, by G. Plekhanov (ed. D. Riazanov, London, 1928), pp. 9 ff.
[2] *De Rer. Nat.* II, 1090.

(Nature, delivered from every haughty Lord
And forthwith free, is seen to do all things
Herself, and through herself of her own accord,
Free of all Gods.)

On the one hand there is the cosmic force which is "responsible" for the vast evolutionary process wherein we form a part, if anything is responsible for it. The modern mind finds the ancient scholastic arguments for the existence of this force or "prime mover" in no way convincing, still less that it partakes of the nature of what we call "mind" or "personality," and even less still that its essence is good. The good seems to arise out of the evolutionary process rather than to have been in it from the beginning. But the good is an immediate datum, and the holiness of good actions is an immediate datum. These are the occasions of modern religion.

From this point of view, the bonds of love and comradeship in human society are analogous to the various forces which hold particles together at the colloidal, crystalline, molecular, and even sub-atomic levels of organisation. The evolutionary process itself supplies us with a criterion of the good. The good is that which contributes most to the social solidarity of organisms having the high degree of organisation which human beings do in fact have. The original sin which prevents us from living as Confucius and Jesus enjoined[1] is recognisable as the remnants in us of features suitable to lower levels of social organisation; anti-social now. If such an idea is accepted, the insistence that we must have some extra-natural criterion of ethical values ceases to have any point. The kind of behaviour which has furthered man's social evolution in the past can be seen very well by viewing human history; and the great ethical teachers, from Confucius onwards, have shown us, in general terms, how men may live together in harmony, employing their several talents to the general good. Perfect social order, the reign of justice and love, the *Regnum Dei* of the theologians, the "Magnetic Mountain" of the poets, is a long way in the future yet, but we know by now the main ethical principles which will help us to get there, and we can dimly see how these have originated during social and biological evolution. There is no need for perplexity as to whether we ought to call evolution morally

[1] There is, of course, the incidental difficulty of continually modifying the letter of the teaching of the great ethical "mutants" to fit changing techniques and increasing knowledge, without losing their spirit.

admirable or morally offensive; it is surely neither. The good is a category which does not emerge until the human level is reached.[1]

The difficulty about religion is that it cannot be considered apart from organised religion as embodied in institutions.[2] In practice, its effects throughout the world are, in the present social context, largely harmful. How far religion can be transformed without the disappearance of the old vessels is a very disputable matter. The detailed beliefs of the past—verbal inspiration, eternal damnation, magical efficacy of prayer for "particular mercies" (in the old phrase), *ex opere operato* rites, miraculous intervention, ascription of psychological states to God, and so on, *are* of course irrevocably of the past, not of the present or the future. None of them are relevant to true religion. Religion is seen not as a divine revelation, but as a function of human nature, in Julian Huxley's words, as a "peculiar and complicated function, sometimes noble, sometimes hateful, sometimes intensely valuable, sometimes a bar to individual or social progress, but no more and no less a function of human nature than fighting or falling in love, than law or literature."[3] Theology, indeed, comes off badly in our modern survey. In so far as it is a codification of the experiences of religious mysticism it is an attempt to reduce to order what cannot be so reduced. In so far as it is a description of such experiences, it is engaged on the fruitless task of describing the indescribable. And in so far as it is occupied with cosmology, anthropology, and history, it is trespassing on legitimate fields of scientific activity.

Many students of these problems at the present time see that the essence of religion is the sense of the holy (Julian Huxley,[4] J. M. Murry,[5] Canon J. M. Wilson and others). Religion thus becomes no more and no less than the reaction of the human spirit to the facts of human destiny and the forces by which it is influenced; and natural piety, or a divination of sacredness in heroic goodness, becomes the primary religious activity. Consider also the following words of one of our most judicious philosophers:—"The identification of this-worldly with material values, other-worldly values alone being recognised as spiritual, is what I am concerned to deny. I maintain that

[1] In this connection C. M. Williams' *Review of the Systems of Ethics founded on the Theory of Evolution* (London, 1893), is still not without value.
[2] Cf. Lenin's remarks on religion in *Works*, Vol. 11, pp. 658 ff. and *Lenin on Religion* (Lawrence, London, n.d.).
[3] J. S. Huxley *What Dare I Think?* (Chatto and Windus, London, 1931), p. 187.
[4] loc. cit.
[5] J. M. Murry, many articles in the Adelphi, and especially 1932, 3, 267.

spiritual good and evil are to be found in the daily intercourse of men with one another in this world, independently of any relation of man to God; further, that the significance of spiritual value does not depend on God or upon the continuance of human beings after the death of the body."[1]

These opinions are not indeed very different from those of many modernist and liberal theologians. The difficulty about religion within the framework of organised christianity is that the "plain statements in Bible and Prayer-Book stand uncorrected and unannotated," so that for simple people they mean what they say. For liberal intellectuals, this may be myth, that may be symbol, this may be a valiant attempt to express the inexpressible, that may be an unfortunate inexactitude due to historical causes—but for the majority of people, everything must be taken literally or not at all. Critics, then, have no alternative but to stand outside the traditional Church and give it advice from a distance, so that their remarks acquire a remote and impractical character. But an acquaintance with the life of religion from the inside convinces one that the sense of the holy cannot be ordered about at will, unhooked from one thing and hooked on to something else, or simply detached from ancient traditions and poured into the cold vacuum of our modern mechanical world. The poetic words of the Liturgy, for instance, philosophically meaningless though they may be, cannot be separated from the numinous feeling which has grown up with them. Though built upon the basis of a world-view which we can no longer accept to-day, they retain, for some of us, enough symbolism of what we *do* believe, to make them of overwhelming poetic value[2].

The upshot of the matter is, therefore, that in practice those who can successfully combine traditional religious life with the life of social and political action appropriate to our time, will be relatively few. It is no good being in a hurry to descry and to welcome the new forms of social emotion; they will emerge in their own good time and perhaps we shall not live to see them. But meanwhile, like the last Pontifex Maximus in Rome,[3] we shall continue those ancient rites which still have meaning for us, while nevertheless being on the best of terms with the clergy and people of the New Dispensation.

[1] Susan Stebbing, *Ideals and Illusions* (London, 1940), p. 31.
[2] Cf. Stewart D. Headlam's *The Service of Humanity* (London, 1882) and *The Meaning of the Mass* (London, 1905).
[3] Or the last priest of Zeus in Richard Garnett's story, *The Twilight of the Gods*. First published 1888, now in Thinker's Library Edition No. 81 (Watts, London, 1940).

Few would wish to maintain to-day that the organised religion of christianity has any gift of temporal immortality, and that it will not find its end just as the religions of ancient Egypt, or of Mexico, found theirs. But some would certainly wish to maintain that religion, as a natural department of the human spirit, has survived these changes and will always survive them. It could also be held that no historic religious system has failed to contribute some element of advance to man's social consciousness. The hope of making religion philosophically respectable is probably quite vain, and the sense of the holy in its ancient form cannot flourish in pure isolation away from its ancient trellis. But will not christian feeling be succeeded by another form of numinous feeling; a new development of social emotions? Even to ask this question is to ask where it could come from. We may be certain that it will not come from the lecture-rooms of academic philosophy, still less from the armchairs of literary critics or the speculations of scientific workers interested in religion from the outside. Will it not come from the factory? Obviously not the factory as we know it to-day, but the factory of the future, the factory of co-operating producers, when the whole system of commercial exploitation has been completely destroyed, and the means of production have been taken over in communal ownership. The most appalling struggles may well be involved in the death-throes of the present system, and we may perhaps expect that the numinous feeling of the future will take its origin from the consequent stress and strain. Is not Mayakovsky's poetry, are not the "Twelve" of Alexander Blok, the symbols of this? But meanwhile, Religion is still resident in her traditional house, and those who would seek her successfully must seek her there as well as in the leaflet distribution and the Trade Union Hall. Auden's words express what is going on:—

> "Love, loath to enter
> The suffering winter,
> Still willing to rejoice
> With the unbroken voice
> At the precocious charm
> Blithe in the dream
> Afraid to wake, afraid
> To doubt one term
> Of summer's perfect fraud,
> Enter and suffer
> Within the quarrel

> Be most at home,
> Among the sterile, prove
> Your vigours, love.

Those of us who have loved the habitation of God's house and the place where his honour dwells, would be well content if the traditional forms of rite and liturgy could survive the coming storm.[1] We would like to fill the old bottles of Catholic doctrine with new wine. The words of the Fathers on equality and social righteousness seem more likely to be fulfilled than we had hoped. But if this revivification of the ancient faith cannot be accomplished, then we shall accept the judgment with a *Nunc dimittis*; those who love both the spirit and the letter will not complain if the spirit be taken and the letter left.

Before leaving the question of the possible forms which numinous feeling may take in future ages, a word should be said of the part which dramatic representations are likely to play. Cinema films of great power (such as Eisenstein's celebrated "Cruiser Potemkin") and also many documentary films portraying the natural and normal life of mankind in its struggle against the external world and its attainment of inner solidarity (such as the "Night Mail" of Grierson and Auden), generate in those who see them emotions to which it would be dangerous to refuse the term "numinous." After all, the religious origins of drama are well known, and it is surely significant that in the Soviet Union, the first great socialist state the world has ever seen, drama, poetry, and all cognate arts flourish as never before.[2] The catharsis of tragedy is only an extreme form of the effect upon individual human beings which any dramatic representation based on fundamental common human values must necessarily have. As the following interesting passage shows, religious exhortations in the old sense will not be needed in the future to awaken men to a sense of their social duty:

> "In one of the novels of Ilya Ehrenburg there is a description of a play given by a travelling company at a collective farm somewhere in northern Russia. Othello was to be played, and the actress who was to take the part of Desdemona (the only

[1] Cf. what George Tyrrell said: "Houtin and Loisy are right; the christianity of the future will consist of mysticism and love, and possibly the Eucharist in its primitive form as its outward bond" (*Autobiography and Life*, London, 1912, vol. 2, p. 377).

[2] Cf. the article on the theatre in the Soviet Union by Herbert Marshall (University Forward, 1941, 7, 10).

sophisticated person present) felt that it was rather absurd. The collective farm had its usual anxieties; the cows were giving only half the amount of milk expected, the ploughing was backward, the fields of mangold-wurzels (or whatever it was) were covered with weeds and badly hoed. In the company's repertory there were Soviet plays, but the provincial actor-manager wanted to strut and gesticulate in the role of the jealous Moor, and nothing else would do. The play began by being misunderstood, but ended with great general emotion in the midst of tears and enthusiastic applause. The peasants of the collective farm were especially affected by Desdemona, but after the performance they amazed her, and even made her cry, because, instead of congratulating her on her acting, they made all kinds of unexpected promises about the augmentation of the milk yield, the improved cultivation of the fields, and the attainment of more than their scheduled production."[1]

Thus even farm labourers, at a comparatively low level of education and culture, could pass over very well from the emotion generated by the tragic situation of individuals, to that involved in the common situation of humanity. Not a few marxist thinkers have, as a matter of fact, foreseen the replacement of organised religion by the arts and drama.[2] This is possible not because the numinous is identical with the aesthetic, but because the never-ending tale of human relationships in the successive stages of social evolution and progress, can itself be the bearer of the numinous. As Feuerbach would have said, man will in this way realise that in the ideological structures of the traditional religions, he was really looking at himself and his own fate and the fate of his society.

Enemies of Human Experience.

There is a kind of fundamental validity attaching to the five great realms of human experience, philosophy, history, science, art, and religion. Each of these has its enemies—those who go about to deny their validity, or their right to exist, or their right to play the part which they do play in our civilisation or our individual lives. Let us consider some of these factors in relation to our main theme.

Against Philosophy come many opponents. Particularly, the

[1] From an essay by a Polish writer, Andrzej Stawar, which a Polish friend of mine and I translated together (Scrutiny, 1937, 6, 21).
[2] Especially G. Plekhanov, *Fundamental Problems of Marxism* (London, n.d.), p. 143.

mathematical logicians point out to us, that there are few, perhaps no, metaphysical propositions which can be translated into the exact language of mathematical logic. Philosophy on this view is an art, a sort of music gone wrong. Among these opponents, however, marxist ethics and orthodox theology cannot be numbered. They, at least, cannot be accused of undervaluing philosophy.

Against Science come many influences, some of which are equally opposed to philosophy. The whole anti-intellectualist movement, so protean in its manifestations in our time, acts in this direction. From the mystical point of view represented by D. H. Lawrence and his followers at one end to the folky-brutal atmosphere of nazism at the other, we have a thoroughly anti-scientific front. For these minds, if so they can be called, scientific internationalism is an illusion, racial factors dominate human actions, and true patriots must think with their blood. Nothing could be more valuable for the armament manufacturers than these views; nothing could be more in line with the feudal vestiges which have for centuries lingered on in the army-officer class. We are witnessing at the present day a wholesale frustration of science.[1] To the capitalist, scientific research is useful, but only relatively in comparison with other and perhaps even cheaper ways of obtaining profits. It is only when these fail that the capitalist now needs the scientist. Again, the conditions of profit-making forbid the introduction of safety measures and the application of labour-saving devices which could greatly increase world-production, while at the same time equalising leisure in the form of a five-hour day under a planned socialist system. Or improved technical methods may be used for actually destroying a part of the produced material, such as coffee or rubber. Or the area of land sown may be compulsorily restricted. Worst of all, perhaps, is the continuing and increasing use of science in war preparations; the development and application of the most diverse scientific researches to rendering the killing of individuals more effective, cheaper, and possible on a still larger scale than ever before. "It does not need much economic knowledge," writes Bernal,[2] "to see that a system of which the essential basis is production for profit, leads by its own impetus into the present highly unstable and dangerous economic and political situation, where

---

[1] See the book of essays, *The Frustration of Science*, by Sir Daniel Hall, J. D. Bernal, J. G. Crowther, E. Charles, V. H. Mottram, P. Gorer, and B. Woolf (Allen & Unwin, London, 1934).

[2] Bernal, J. D., "National Scientific Research," Progress, 1934, **2**, 364.

plenty and poverty, the desire for peace and the preparation for war, exist side by side; but it does require far more knowledge to see how an alternative system could be built up. And yet, unless scientists are prepared to study this they must accept the present state of affairs and see the results of their own work inadequately utilised to-day and dangerously abused in the near future." Thus the figures of the annual government grants speak for themselves. In 1933, for example, the Medical Research Council received £139,000 and the Department for Scientific and Industrial Research £443,838, while the research grants for the Army, Navy, and Air Force together were £2,759,000, i.e. five times as much as the whole total for civil research.[1]

Another of the influences working in our time against science is the outcome of modern psychology.[2] An argument nowadays need not be answered; it is sufficient to trace it back to the previous psychological history, and hence the prejudices, of the person who propounds it. A misunderstanding of marxism, with its insistence upon the class basis of science, has exposed it to this accusation, but it is perfectly legitimate to apply the class theory of history to the history of science, and the results are frequently highly convincing. On the other hand the fascist struggle (especially in Germany) against "objective science," based on the racial theory of history, which has no scientific basis of any sort, is the most dangerous form of this kind of attack which exists, though it can only be seriously proclaimed to the masses under conditions where all criticism is silenced by state power. As for Art, it does not pay.[3] No further enemy is needed. And History, as eminent capitalists have assured us, is bunk.[4]

Against Religion come so many forces that it is hard to count them. The general trend from religion to science which took place in the Hellenistic age and the late Roman Empire repeated itself again in our own western European civilisation from the Renaissance onwards. Religion has had to face the great pretensions of the mediaeval secular power, the mechanical philosophies of the 17th century, the enlightened atheism of the 18th century, and the Victorian agnosticism of the last age. Bourgeois agnostics and proletarian atheists have attacked it from all sides. It is surprising that there is anything left of it: and few people seem even to know what it is. Thus an

[1] Budget Estimates.
[2] Cf. Joad, C. E. M., *Guide to Modern Thought* (London, 1933); and *Under the Fifth Rib: An Autobiography* (London, 1932).
[3] See p. 138.
[4] The dictum is attributed to Mr. Henry Ford.

anonymous writer recently began an article on agnosticism with the words: "The essence of religion is faith, the ability to accept as a truth a hypothesis for which there is no positive evidence."[1] Or again, in *Moscow Dialogues*, Socratov says,[2] "We are rather at a loss to point to anything of a positive character in religion. If you can suggest anything positive, I shall be glad to hear it"; and the Bishop (very conveniently) replies, "Well, first of all, the Church has always stood, even in its darkest days, for law and order." The first of these writers was confusing, as is so common, theology with religion. Theology has to accept hypotheses for which there is no positive evidence, because in a system so unlikely as the universe, of which there is only one, no comparisons can be made by which to test the credibility of anything. This is no argument in favour of theology, which may or may not be a necessary evil, but on the other hand, it does not discredit religion. The second was erecting an episcopal man of straw in order to have the pleasure of hearing the opium-merchant give himself away red-handed. But the statement is not historically true; when Irenaeus, Clement, and Tertullian were alive; when Lilburne, Rainborough, and their "russet-coated captains" were riding; the Church was not on the side of law and order. Christians were able to imagine a better law and a juster order than the established system of the Roman Empire,[3] or the government of that "Man of Blood," King Charles I.

The clearest understanding of religion has been given, in my view, by the work of Rudolf Otto,[4] a German theologian, who described it as the sense of the holy. In primitive communities we see this "numinous sense" applied to all kinds of worthless objects and rites, and later incorporated in the apparatus of State government, but in the great religions of the world it forms the essential backbone of the experience of their participators. In christianity, where the ethic of love found its greatest prophets, the numinous sense has become attached to the highest conception of the relations between man and man that we know. The christian who becomes a communist does so precisely because he sees no other body of people in the world of our time who are concerned to put Christ's commands into literal

[1] New Statesman, 1934, **8**, 332 (September 15th).
[2] Hecker, J., *Moscow Dialogues* (Chapman & Hall, London, 1933), p. 191.
[3] On the socialism of the Apostolic Fathers, see the essay of Charles Marson in the collective work by Tom Mann and others, *Vox Clamantium*, ed. Andrew Reid (London, 1894); and also his *God's Co-operative Society* (Longmans, Green, London, 1914).
[4] See especially Otto, R., *The Idea of the Holy* (Oxford, 1923).

execution. If for seventeen centuries the Church has tended to put allegorical constructions on the Gospels, we know that the christians of the first two centuries did not do so.

That religion has been, and largely is, "the opium of the people" is plainly undeniable. Proletarian misery in this world has been constantly lightened by promises of comfort and blessedness in the world beyond the grave, an exhortation which might come well enough from some ecclesiastical ascetic who did not spare himself, but very ill indeed from the employer of labour or the representative of the propertied classes. But the conclusion usually drawn, namely, that religion could have no place in a socialist State, where no class-distinctions existed, does not seem to follow. Because religion has been often used as a social opiate in the past, there seems no reason why this should be so in the future. "Religion would continue to exist," writes A. L. Rowse,[1] "in the socialist community, but on its own strength. It would not have the bias of the State exerted in its favour, as it has had so strongly in England up to the present, and in greater or lesser degrees in all western countries." It may indeed be said that religion is "the protest of the oppressed creature,"[2] and that therefore when social oppression, in the form of the class-stratified society, is done away with, the private need for religion will vanish as well as the class which profited by it. This, however, is to forget what we could call "cosmic oppression," or creatureliness, the unescapable inclusion of man in space-time, subject to pain, sorrow, sadness and death. Shall we substitute for the opium of religion an opium of science? It has always been the tacit conviction of the social reformer and the person occupied with the practical application of scientific knowledge that by man's own efforts, not merely minor evils, but the major evils of existence may be overcome. This is expressed in that great sentence: "Philosophers have talked about the universe enough; the time has come to change it."[3] But the problem of evil is not capable of so simple a resolution. So long as time continues, so long as change and decay are around us and in us, so long will sorrow and tragedy be with us.[4] "Life is a sad composition," as

[1] Rowse, A. L., *Politics and the Younger Generation* (Faber, London, 1931), p. 194.
[2] Marx, K., *Introduction to a Critique of Hegel's Philosophy of Law*.
[3] And also in the great concluding paragraph of John Stuart Mill's *On Liberty* (written between 1854 and 1859).
[4] Cf. Kierkegaard's distinction between "tribulations" (natural troubles which can only be endured) and "temptations" (troubles due to, and soluble by, acts of will), discussed by Auden in *New Year Letter*, 1941, p. 132.

Sir Thomas Browne said, "we live with death and die not in a moment." Or, in the words of the *Contakion*, "For so thou didst ordain when thou createdst us, saying, 'Dust thou art, and unto dust shalt thou return'; wherefore all we who go down into the grave make our song unto thee, sighing and saying, Give rest, O Christ, to thy servants with thy saints, where sorrow and sighing are no more, neither pain, but life everlasting." The whole realm of thought and feeling embodied in these phrases is fundamentally natural and proper to man, and there is little to be gained by trying to replace it by a eupeptic opium, derived from too bright an estimate of the possibilities of scientific knowledge. Driven out, it will return in the end with redoubled force.

Fundamentally natural and proper to man, the sense of the holy is as appropriate to him as the sense of beauty. As we have seen, the "moral theology" of communism lacks a doctrine of God, but this does not affect the existence of the sense of the holy. After all, the theology of the Gospels was not very complicated—Jesus did not meet disease and hunger by persuading people that blessedness was already theirs if they would accept a dogmatic intellectual system; but by curing sickness and distributing bread. This was the practical aspect of his teaching about love. In the motives of atheist communists we detect, therefore, that which is worthy of numinous respect, for they are working to bring in the World Co-operative Commonwealth.

Those who deny the importance of the sense of the holy are in an analogous position to those who cannot appreciate music or painting. It is an attitude towards the universe, an attitude almost of respect, for which nothing can be substituted. "The problem of death," it has been said,[1] "is not a 'problem' at all, it is due simply to the clash between an idealistic egoistic philosophy and the disappearance of the individual, not in the least to the fact of death." On this epicurean view, science reveals facts to us so clearly as to reconcile us to them.[2] But it is not our own death that we are thinking of. We may well be content to live on only in the effects which our living has produced on our generation and those that come after.[3] The point is that no matter now much we know in the classless State about the biology of death, we shall still suffer when someone that we have loved

[1] Pascal, R., *Outpost*, 1932, **1**, 70.
[2] All sciences have as their aim the transformation of tribulations into temptations, Auden, loc. cit. But the process is asymptotic.
[3] A point of view admirably put in Afinogenov's play *Distant Point*.

suddenly dies or is killed. The question then reduces itself to a matter of taste; shall we bury him with unloving haste and a callous reference to the unimportance of the individual? Or shall we remember, as we fulfil the rites of a liturgical requiem, that this is the common end of all the sons of men, and so unite ourselves with the blessed company of all faithful people, those who earnestly looked and worked in their generation for the coming of the Kingdom? It is true indeed, as Merejkovsky has said, that whether we believe in Christ or not, we must certainly suffer with him. And, indeed, it is my opinion that if the ancient christian modes of satisfying this numinous sense are discontinued (Eliot's *"vieilles usines désaffectés"*), other liturgical forms will be devised to play their part in attempting to express that which cannot be expressed. This we already see in such cases as the tomb of Lenin himself, and the Red Corners.

In the Timiriazev Institute at Moscow I examined the red banner which the scientists there were accustomed to carry on May Day and other public occasions. It was of a velvety cloth with yellow fringes and an elaborate hammer and sickle. How tawdry some of our respectable middle class people in England would have thought it. How they despise the native decorations and pictures on British Trade Union banners on the rare occasions when these pass through the streets in the light of day. But to me it was quite clearly numinous, one with the cross of our salvation and the *Vexilla Regis* indeed, conspicuous in the vanguard of humanity moving from the captivity of necessity into the glorious liberty of the children of God. But is this process ever complete? Are we not all for ever in bondage to space-time? Is not this bondage our final evil? It is absurd to say that "with the denial of an objective creator, socialism forgets the problem of evil." Certainly no "person" is now responsible, but in whatever society man arranges himself he must take up some attitude towards the universe, and to the fate of individuals in it, and in this attitude, the sense of the holy will always be an element.

Scientific Opium.

Not to be awake to the iniquity of class oppression, then, is religious opium. Scientific opium would mean not being awake to the tragic side of life, to the numinous elements of the world and of human effort in the world, to religious worship. Scientific opium has often been thought an integral part of marxism by its opponents, but for us the question is what break with tradition the contribution of

England and the west of Europe to the socialisation of the world must involve. In this connection there are two considerations which seem relevant, but which have not often been discussed. In the first place, it is a historical coincidence that the early marxists adopted the anthropological and psychological arguments against religion which were fashionable at the time. These arguments are insufficient ground for condemning one of the greatest forms of human experience. The anthropological arguments all confuse origin with value, as if primitive barbarism were not in the end responsible for science, art, and literature just as much as for religion. To say that the concept of God is derived from, or modelled on, the relation of primitive exploiting lord to primitive exploited slave is to say nothing about the religious value of the concept in a society where exploitation has been abolished (should it continue to exist there); still less about religion itself as opposed to theology or philosophy. For religion does not know what God is; it only knows him if he exists to be worthy of worship—a God comprehended would be no God—and it does not know why the universe is as it is, but only that there is holiness in it. An excess of mystical religion may indeed engender an attitude of inactivity against the external world, but we need it as a salt, not a whole diet. Must we have prohibition in the classless State because some men drink too deeply to-day? In the end, there is but one end, and communism can overcome the last enemy no more than any other of man's devices. It is difficult, no doubt, to combine scientific "pride" with religious "humility," but the best things often are difficult.

In the second place, the Byzantine nature of eastern christianity is relevant. From the very beginning the Byzantine Church showed a speculative rather than a practical tendency.[1] The east enacted creeds, the west discipline. The first decree of an eastern council was to determine the relations of the Godhead; the first decree of the Bishop of Rome was to prohibit the marriage of the clergy. Eastern theology was rhetorical in form and based on philosophy; western theology was logical in form and based on law. The Byzantine divine succeeded to the place of the Greek sophist; the Latin divine to the place of the Roman advocate. The eastern Church, therefore, occupied with philosophy and theology, made little or no pretensions to control of economic affairs, no attempt to subordinate the secular power to

[1] See Milman, H. H., *History of Latin Christianity* (Murray, London, 1867); and Stanley, A. P., *Lectures on the History of the Eastern Church* (Dent, Everyman edition).

itself in the interests of a particular theory as to how the mercantile life should be lived. The Patriarchs, chosen from a monastic order remarkable for its detachment from secular business, left all economic questions to the chamberlains and officials who thronged the imperial court. After the fall of Byzantium, this same tradition of complete other-worldliness transferred itself to the Church of Russia. The Russian Orthodox Church had no pope, no Hildebrand, to impose a theological system of economics on Russian society. It had no scholastic philosophers, no "mediaeval clerks" to dictate to kings and rulers what measures they should take to secure social justice. It had nothing corresponding to our 17th-century High Churchmen, or to our 19th-century Anglo-Catholics, reviving those traditions and reminding men of the ideals of a pre-capitalist age. When capitalism, in the time of Peter the Great, reached Russia, it found a perfectly virgin soil for its operations, and had no such uphill task as it found in the west. In three generations it enslaved a population which could make no appeal to any distinctively christian social theories. The appeal would have been vain, for the Orthodox Church had no such theories, and had never developed the first beginnings of them. On the contrary, it had become completely identified with the process of exploitation of the Russian people. The contrast between this situation and our own is quite remarkable.

It may be said that the meaning of the phrase "religious opium" was that by anaesthetising the people, it prevented them from performing those social actions necessary for social progress, combining in unions, rebelling against exploitation, fighting the possessing class in every possible way. "Scientific opium" could have no such meaning. Yet I think it has, and it may be explained as follows. It is a blindness to the suffering of others. A certain degree of ruthlessness is absolutely inevitable in the period of revolutions when the people are defending themselves against the final attack of the possessing class which sees itself on the verge of expropriation. "Revolutions," said Lenin, "cannot be made without breaking heads." But just as Lunacharsky (whose role will be better appreciated by later historians) pleaded successfully for the preservation of certain buildings, art treasures, etc. in the heat of the revolution; so it is always necessary for the christian man (even he who without reservation allies himself with the revolution) to plead for the retention of certain christian principles in dealing with people. The ruthlessness necessary in a revolutionary period or an age of wars may too easily pass over, especially in a society based on science, and the more so the more it is so based, into a ruthlessness derived from the very statistical character of the scientific method itself. The ruthlessness with which a biologist throws out an anomalous embryo useless for his immediate purpose, the ruthlessness with which an astronomer rejects an aberrant observation, may too easily be applied to human misfits and deviationists in the socialist world order. The witness of the christian man may then recall the marxist to a sense of the fundamentally unmarxist character of such treatment. It is unmarxist because no philosophy recognises the emergence of levels of high organisation better than dialectical materialism, and the individuals of which the human social collectivity is built up are themselves the most complicated organisms in the living world.[1] Hence christian love in the form of tolerance is transformed into a recognition of the manifold forms which human thought and being may take. As long as aberrant individuals are not permitted to be a danger to the socialist state, the greatest tolerance should prevail. There is no need for marxists to follow the example of those many unchristian christians who manned the Inquisition, the witch-hunting tribunals, and the boards of godly divines in Geneva, Westminster and Massachusetts.

We have here a principle of genuine importance. Christian theology has been called "the grandmother of bolshevism," since communist planning alone has seen how to incorporate the love of one's fellow-men in the actual structure of economic life. Some have seen another ancestor in the rationalist and philanthropic ethic of ancient confucianism. But communism is based just as much on the findings of natural science and the method of science itself. The socialist society must therefore guard against taking over from science too much of scientific abstraction, scientific statistical ruthlessness, and scientific detachment from the individual.

Christian Theology the Grandmother of Bolshevism.

Important for the decay of religion in our time is the general and increasing domination of the scientific mind, or, rather, of a popular version of the state of mind characteristic of the scientific worker. Constantly growing power over external nature leads to a tacit belief in the possibility of solving the problem of evil by what might almost be called a matter of engineering. The principle of abstraction leads to a weakening of that attention to the individual and the unique

[1] This explains Blake's antipathy to Newton.

which must always be an integral part of the religious outlook. The principle of ethical neutrality leads to a general chaos in the traditional systems of morals, and hence to decay in the religious emotion formerly attached to the performance of certain actions. The emphasis laid by the scientific mentality on the quantitative aspects of nature runs diametrically counter to the emphasis which religion would like to lay on the other aspects of the universe. And, above all, in actively interfering with the external world, in persistently probing its darkest corners, science destroys that feeling of creaturely dependence upon, and intimate relation to, a transcendent and supernal Being, which has certainly been one of the most marked characteristics of the religious spirit. In the modern world, Epicurus and Lucretius have come into their own.

But here we find, paradoxically enough, that communism and the christian religion are again on the same side. If these effects of the domination of science were to operate alone, we should have a truly soulless society, much as is depicted by Bertrand Russell in *The Scientific Outlook*,[1] and by Aldous Huxley, satirically, in *Brave New World*.[2] This is what we shall certainly get if capitalism can establish itself anew and overcome the forces of fanatical nationalism which threaten to disrupt it. For capitalism has a fundamentally cheap estimate of the value of human life; mine disasters and wars alike are but passing incidents in a society where the only principle recognised is that might is right. Communism and christianity, on the contrary, estimate life highly. Ultimately the distinction here resolves itself into what kind of human society we wish to aim at, and the choice may be in a sense aesthetic. The logical continuation of the capitalist order would be the tightening and stabilisation of class-stratification, which seems to be the essential function of fascism. This could then, in time, be further fixed as biological engineering becomes more powerful. In such a civilisation, the Utopia of the bourgeoisie, where an abundance of docile workers of very limited intelligence was available, the class stratification would be absolute, and the governing class alone would be capable of living anything approaching a full life.[3] Biological engineering would have done what mechanical engineering had failed

[1] (Allen & Unwin, London, 1931.)
[2] (Chatto & Windus, London, 1932.) We shall analyse this book in what follows.
[3] It is of much interest that the similarity between fascism and the ancient caste-system of India is expressly admitted in *Sanatana Dharma*, an advanced textbook of Hindu religion and ethics, published by the Central Hindu College, Benares, 1923, pp. 240 ff. Both are said to be based on the doctrine of immortality.

to do, and flesh and blood would have been adapted to machinery rather than machinery to flesh and blood. Nevertheless the converse process is equally possible, i.e. a continually increasing automatism of machine operations, and hence an increasing liberation of man from the necessity of productive labour. With the increase of leisure would come an enormous increase in the beneficial and pleasurable occupations available for the workers. This is what is meant by the readiness to sacrifice the bourgeois liberty of to-day for the much greater liberty of the classless State. And these two alternatives are even now offering themselves to us, with capitalism on the one side and christianity and communism together on the other. It is a pity that Spengler's aphorism is not more widely known: "Christian theology is the grandmother of bolshevism."

"Utopias," wrote Berdyaev, in a passage which Aldous Huxley chose for the motto of *Brave New World*," appear to be much more realisable than we used to think. We are finding ourselves face to face with a far more awful question—how can we avoid their actualisation? And perhaps a new period is beginning, a period when intelligent men will be wondering how they can avoid these Utopias, and return to a society non-Utopian, less perfect but more free." Huxley's book was a brilliant commentary on this.

His theme is twofold, one of its aspects being the power of autocratic dictatorship, and the other, the possibilities of this power, given the resources of a really advanced biological engineering. The book opens with a long description of a human embryo factory, where the eggs emitted by carefully-tended ovaries are brought up in their development by mass-production methods on an endless conveyor belt moving very slowly until at last the infants are "decanted" one by one into a remarkable world. The methods of education by continual suggestion and all the possibilities of conditional reflexes are brilliantly described, and we see a world where art and religion no longer exist, but in which an absolutely stable form of society has been achieved, first by sorting out the eggs into groups of known inherited characteristics and then setting each group when adult to do the work for which it is fitted; and secondly by allowing unlimited sexual life (of course, sterile). This idea was based on the suggestion of Kyrle[1] that social discontent, which has always been an important driving force in social change, is a manifestation of the Oedipus complexes of the members of society and cannot be removed by

[1] R. M. Kyrle, Psyche, 1931, **11**, 48.

economic means. With decrease of sexual tabus, these psychologists suggest, there would be a decrease of frustration and hence of that aggressiveness which finds its sublimation in religion or its outlet in political activity. Thus in the society pictured by Aldous Huxley, erotic play of children is encouraged rather than prevented, universal but superficial sex relations are the rule, and indeed any sign of the beginning of more deep and lasting affection is stamped out as being anti-social.

Perhaps only biologists really appreciated the full force of *Brave New World*. They knew that Huxley included nothing in his book but what might be considered legitimate extrapolations from already existing knowledge and power. Successful experiments are even now being made in the cultivation of embryos of small mammals *in vitro*. One of the most horrible of Huxley's predictions, the production of numerous workers of low-grade intelligence and precisely identical genetic constitution from one egg, is theoretically quite possible. Armadillos, parasitic insects, and even sea-urchins, if treated in the right manner, will "bud" in this way now, and the difficulties in the way of effecting it with mammalian and therefore human eggs are probably purely technical.

It is just the same in the realm of philosophy. There are already among us tendencies leading in the direction of Huxley's realm of Antichrist. Fascism seeks no justification other than existence and force. Its philosophy is one in which there is no place for science. Science ceases to be the groundwork of philosophy, and becomes nothing but the mythology accompanying a technique. Divorced from religion, ethics and art, as well as from philosophy, it proceeds to do the will of wicked and ungodly rulers upon humanity. "The scientific society in its pure form," as Bertrand Russell has said, "is incompatible with the pursuit of truth, with love, with art, with spontaneous delight, with every ideal that men have hitherto cherished, save only possibly ascetic renunciation. It is not knowledge that is the source of these dangers. Knowledge is good and ignorance is evil—to this principle the lover of the world can admit no exception. Nor is it power in and for itself that is the source of danger. What is dangerous is power wielded for the sake of power, not power wielded for the sake of genuine good."

This train of thought leads us finally to consider on what ground communism can stand as against nietszchianism or other doctrines of the "superman." These may be, for all we know, perennial, if they derive primarily from specific psychological types, and may appear long after the classless society has been established. Thus if it be claimed that the fulfilment of the personality of one sort of individual necessitates the injury or exploitation of others, on what ground does communist theory refute the claim? The ethical superiority of social equality is in fact at issue. Barbara Wootton[1] well points out that "every type of economic organisation will turn top-heavy unless it is quite definitely and deliberately weighted in favour of the weak, the unfortunate, and the incompetent." What justification can there be for this, except the ἀγαπή τοῦ πλησίον of the Gospels, one of the two commandments on which hang all the Law and the Prophets? And this leads us to ask whence came the noble hatred of oppression found in Marx, and whence arises this passion in all the communist confessors and martyrs of the present century? It cannot be a coincidence that marxist morality grew up in the bosom of christianity after eighteen christian centuries, as if the phoenix of the Kingdom should arise from the ashes of the Church's failure.

---

[1] *Plan or No Plan* (Gollancz, London, 1934), p. 106.

Laud, the Levellers, and the Virtuosi

(A contribution to the book of essays, *Christianity and the Social Revolution*, 1935)

Seventeenth-Century England.

IN seventeenth-century England, we have the fascinating picture of a balance trembling on a poise of equal weights—western religion having lost little of its ancient power, western science having gained its first magnificent victories. Even within religion there was a moment of equal poise between the antagonists, before the mediaeval tradition, in the form of the Church of England, ceded the power to the Protestant and Puritan bodies. This contrast, like the former, was but an aspect of what was perhaps a more deep-seated one, namely, the passing of power from the feudal aristocratic and monastic system to the middle or bourgeois class arising out of the mediaeval town merchants. The civil war and the Commonwealth were the outward and visible signs of the victory of this new order. The abolition of the laws against usury; the "freeing" of trade from galling restrictions; the beginning of large-scale industrial "ventures"; the great advances in technology backed by science; the complete removal, in a word, of mercantile and economic life from theological control—all signified the triumph of the middle class.

The Laudian Divines.

Of all the ages of the Church's history after the first two centuries there are few which can compare in brightness with the Church of England in the seventeenth century. Poets like George Herbert, Richard Crashaw, Henry Vaughan, Henry King, and, we might add, Thomas Browne and Jeremy Taylor; saints like Nicholas Ferrar and Thomas Ken; careful restorers of what was destroyed, like John Hacket, Matthew Wren and John Cosin; scholars like Lancelot Andrewes and Henry More, statesmen like William Laud—all combined to give the period a charm and depth which can never be forgotten by those who have studied it.[1] But while we usually

[1] Cf. Grierson, H. J. C., *Cross Currents in English Literature of the Seventeenth Century* (Chatto & Windus, London, 1929); and Willey, B., *The Seventeenth-Century Background* (Chatto & Windus, London, 1934).

think of these men in connection with their importance in the history of theology or philosophy, or with regard to the literary beauty they created, we forget that there was a significant economic aspect to their existence. This may be summarised by saying that they were the representatives of the old conceptions of social justice in economic affairs, and were opposed to the new aims of capitalist freedom in commerce.

William Laud, Archbishop of Canterbury, usually appears in history as the instrument of monarchical oppression, and not as the champion of popular agrarian rights. Yet there is no doubt that among the economic causes of the civil war and of Laud's own fall was the opposition which he aroused among landowners by his agrarian policy.[1] The problem of enclosures was by no means new in English economic life in the days of the Stuarts; it runs, indeed, like a connecting thread through all the economic life of the country from the early Middle Ages to the nineteenth century.[2] Pasture was more of a business proposition than tillage; it was capitalist in its methods, and offered a better chance of big profits. It was the basis of the great late mediaeval wool trade. But the social obligations of the feudal landowner were forgotten; the peasant became a landless and insecure wage-worker, and the land came to be looked upon solely as a source of profit.

To whom were the peasants to turn for redress? Not to the justices of the peace, for these were of the landowning class; not to Parliament, where the same interests reigned. They most commonly appealed to the King's Ministers, the Privy Council, and the Church. Laud strove by every means in his power to prevent such enclosures as depopulated the countryside, and, by heading the Commission on Depopulation, infuriated the capitalist landowners whose interests were aligning them with the industrial capitalists of the towns. Laud had no respect for persons, and would allow no man, however powerful, to transgress what he called the common law of Christ, binding upon man as man. Peter Heylyn, his chaplain and biographer,[3] seems to have thought that Laud could have kept his place and saved his life if he

[1] See Cole, G. D. H., Church Socialist, 1915; Hancock, T., *The Pulpit and the Press, and other Sermons preached at St. Nicholas Cole Abbey* (Brown & Langham, London, 1904); Sykes, N., The Way, 1941, **3**, 18; Schlatter, R. B., *The Social Ideas of Religious Leaders, 1660-1688* (Oxford, 1940).

[2] Cf. the passage on it in Thomas More's *Utopia*.

[3] Heylyn, P., *Cyprianus Anglicus; or the History of the Life and Death of William, Archbishop of Canterbury* (London, 1668).

had paid adulation to the great enclosers, but "he failed in so many necessary civilities to the nobility and gentry" that it was clear he was their enemy and the peasants' friend. His visitation articles, in particular, questioned the churchwardens closely concerning enclosures, detentions, inversions, and so on. To have such questions put to them—as a writer complained a few months before Laud's death, when the Archbishop was safely in the Tower—was a "vassaldrie of the gentry of England," who, from the time of the Tudors, had been impropriating wholesale the common property of the people, their common Church, their common lands, and their common free schools. "Many nobles and worthy gentlemen," said the complainant, "are curbed and tyrannised over by some base clergy of mean parentage." As Clarendon says,[1] "The shame, which they called an insolent triumph over their degree and quality, and a *levelling* them with the common people, was never forgotten, and they watched for revenge."

A final instance: from among his injunctions to the Dean and Chapter of Chichester—"Use some means with Mr. Peter Cox" (a land-grabbing alderman of that city) "that the piece of ground called Campus now in his possession be laid open again, that the scholars of your free school may have liberty to play there, as formerly they had. And if he shall refuse, give us notice, or our vicar-general, upon what reason and ground he does it."

But if some bishops were fighting on the agrarian front, others were leading the struggle against usury.[2] Lancelot Andrewes, the admirable Bishop of Winchester, preached incessantly against it. He made short work of the settlement of 1571, which had legalised the taking of ten per cent. Joseph Hall, Bishop of Norwich; John Jewel, Bishop of Salisbury, and George Downam ("the hammer of usurers") Bishop of Derry, were all prominent in this work.[3] But the merchants persisted ever that "it is not in simple divines to saye what contract is lawfull, and what is not."

In the end these controversies were not settled except by force of arms. In the civil war, the industrial and commercial cities and areas were in general on the parliamentary side; the agricultural parts of the country, except the Eastern Counties, where puritanism

[1] Clarendon, *History of the Rebellion*.
[2] See the classical *Religion and the Rise of Capitalism*, by R. H. Tawney (Murray, London, 1926); and also H. M. Robertson's *Aspects of the Rise of Economic Individualism* (Cambridge, 1933).
[3] Cf. Blaxton, *The English Usurer* (London, 1634).

was so strongly entrenched, were royalist. It was no coincidence that the chaplains of the volunteer regiments of London were Presbyterian, as opposed to the Anglican and Roman Catholic influence on the other side.[1] The golden age of seventeenth-century Anglicanism stood, in fact, in its economic aspect, for a scarcely altered version of mediaeval theocracy. The bishops were "mediaeval clerks," determined to control the market-place. The victories of Cromwell opened the door for the era of capitalist enterprise, and, when the Church of England regained its possessions at the Restoration, it was at the price of most of its militant spirit. In 1692, when one David Jones was so indiscreet as to preach at St. Mary Woolnoth, in Lombard Street, a sermon against usury, his career in London was brought to an abrupt conclusion.

The Levellers.

Let us now cross over to the other side in the civil war in order to trace another movement of great interest—that of the Levellers. If we regard, as we must, the civil war as England's "bourgeois revolution," we should expect to find a certain number of true socialists on the left wing of the revolutionary party, men who would not be content with the political equality which the Cromwellian system would give, but would demand economic equality as well. This is indeed exactly what happened, and from 1647 onwards the parliamentary side was split into two portions, the main body quite satisfied with the defeat of everything that the royalist and anglican forces had stood for, and a smaller body desirous of pushing on towards what we should now call a Socialist State.[2] The fortunes of

[1] It must be understood that names which we use today only for religious denominations had in the seventeenth century a strong political significance. On the Parliamentary side there were few or no Anglicans. Presbyterianism stood for the moderate puritanism of the middle-class merchant interest, and was compatible with compromise as regards the royalist issue until an advanced stage of the Commonwealth. Independency (the predecessor of modern Congregationalism) was adhered to by all the more revolutionary elements, and where the revolution eventually split in taking its inevitable "two steps forward and one step back" was between the so-called "gentlemen-Independents" such as Cromwell, Ireton and the other leaders on the one hand; and the Levellers, Diggers, etc., on the other.

[2] For the Levellers the most accessible book is Henry Holorenshaw's *The Levellers and the English Revolution* (London, 1939); further see the work of E. Bernstein, *Cromwell and Communism; Socialism and Democracy in the Great English Revolution* (Allen & Unwin, London, 1930); also *English Democratic Ideas in the Seventeenth Century* by G. P. Gooch, 2nd edition, edited H. J. Laski (Cambridge, 1927); T. C. Pease, *The Levellers' Movement* (Washington, 1916); D. W. Petegorsky, *Left-Wing Democracy in the English Civil War* (London, 1940); A. S. P. Woodhouse, *Puritanism and Liberty* (London, 1938).

this smaller portion, the Levellers, varied considerably; at one time they were sufficiently strong to take the field against Cromwell's own forces in a short campaign which receives little or no mention in orthodox history books,[1] while towards the end of the Commonwealth they were mostly in exile, reduced to plotting in company with exiled royalists.

As the extreme left wing of the puritan and parliamentary movement, they were, of course, implacably opposed to everything that the Anglican divines stood for. However devoted the Anglican divines might be to ideals of social justice, their positions made them, in the eyes of the revolutionary puritans, pillars of oppression and symbols of the old régime. The moderate puritans were well aware of the situation. In 1641 the poet Edmund Waller, a Presbyterian, said in the House of Commons that though it might be well to restrict episcopacy, it were better not to abolish it altogether. "I look upon episcopacy," he said, "as a counterscarp or outwork; which, if it be taken by this assault of the people, and withal, this mystery once revealed, that we must deny them nothing when they ask it thus in troops, we may, in the next place, have as hard a task to defend our property as we have lately had to recover it from the royal prerogative. If, by multiplying hands and petitions, they prevail for an equality in things ecclesiastical, the next demand perhaps may be the like equality in things temporal." This was the authentic voice of the rising middle class, determined to do away with feudal absolutism, but equally determined to keep the property privilege for itself. Waller went on to say, "I am confident that, whenever an equal division of lands and goods shall be desired, there will be as many places in Scripture found out which seem to favour that, as there are now alleged against the prelacy or preferment of the Church. And as for abuses, when you are now told what this and that poor man hath suffered by the bishops, you may be presented with a thousand instances of poor men that have received hard measure from their landlords." Waller was a keen-sighted man. The preaching troopers who began by finding no warrant in scripture for prelates, ended by finding none there either for the class-domination of temporal rulers.

The Levellers first appear about the year 1647, at which time the victorious army was dividing into the two sections above-mentioned, the "gentlemen-independents" or "Grandees" being opposed to the

[1] Contrast the excellent *History of Feudalism* by A. Gukovsky & O. Trachtenberg (Moscow, 1934).

"honest substantive soldiers" and their elected leaders or "Agitators."[1] In the following year there appeared as one of the numerous pamphlets of the time, an important programme of reforms, *The Agreement of the People*, "to take away all known and burdensome grievances."[2] One of the authors was the indefatigable Lieut.-Col. John Lilburne ("freeborn John").[3] At this time, the Levellers numbered among them many interesting and important pioneers, such as Colonel Rainborough and other officers in the Army and Richard Overton and William Walwyn in civil life. Walwyn wrote little, but he said: "The world will never be well till all things are common." It would not by any means be "such difficulty as men take it to be to alter the course of the world in this thing; a very few diligent and valiant spirits may turn the world upside down if they observe the seasons and shall with life and courage engage accordingly." To the objection that community of property would upset all and every Government, he answered that "there would then be less need of government, for then there would be no thieves, no covetous persons, no deceiving and abuse of one another, and so no need of government. If any difference do fall out, take a cobbler from his seat, or any other tradesman that is an honest and just man, and let him hear the case and determine the same, and then betake himself to his work again." There is a remarkably modern ring about these sentiments. They form a contrast indeed to the attitude of Cromwell, who was always protesting that he was a "gentleman born."

Perhaps the most remarkable pamphlet of the Levellers was *The Light Shining in Buckinghamshire*, which laid down what "honest people desire:—(1) a just portion for each man to live, so that none need to beg or steal for want, but everyone may live comfortable; (2) a just rule for each man to go by, which rule is to be found in Scripture;" (3) equal rights; (4) government judges elected by all

[1] The word (first used at this time) meant a delegate to the Army Council elected by the rank and file of a unit in order to *do* something, i.e. to see that the Council carried out the wishes of the Regiments (for instance, not to disband until the Agreement of the People was accepted and implemented by Parliament.

[2] It demanded annual Parliaments, elected by a universal manhood suffrage, abolition of all coercive laws respecting religion, a national militia recruitment to which should recognise conscientious objections to bearing arms, replacement of all tithes, tolls, etc., by a single income tax, provision of work for the unemployed, an old age pension, and adequate medical provision for all sick persons. In 1648! Parliament condemned the first edition but the Army continued to produce them.

[3] Cf. J. Clayton's biography of him in *Leaders of the People* (Secker, London, 1910). It is interesting that some of his writings have been translated into Russian, ed. V. Semenov (Moscow 1937).

the people; (5) a Commonwealth "after the pattern of the Bible." Here the land was expressly stated to be the property of the whole people, and, as we should say, its "nationalisation" demanded.

In April 1649, while Lilburne and other Leveller leaders were confined in the Tower, there suddenly appeared at Cobham in Surrey a number of men armed with spades, who began to dig up uncultivated common land at the side of St. George's Hill, with the intention of growing corn and other produce.[1] They proposed to prove that "it was an indeniable equity that the common people ought to dig, plow, plant, and dwell upon the commons without hiring them or paying rent to any." A fortnight later they were arrested by two troops of horse, sent down by Cromwell, and their leaders, William Everard and Gerrard Winstanley, brought before him. The examination showed that these "true Levellers," as they called themselves, were in reality trying to found what we should now call a "collective farm," and their conviction was that, when men began to see the success of their venture, they would join it, and so establish in course of time a widespread co-operative system. The beginning was to be on common-land, for which they asked no permission, since from of old it had been the common property of the English people.

Of course, these beginnings were not allowed to proceed far, and though Winstanley succeeded in establishing collective farms at several other places besides Cobham, they were very short-lived. The "true Levellers" seem eventually to have joined other later movements, such as that of the Quakers, which had a more other-worldly background. Winstanley produced a book, however (*The Law of Freedom in a Platform, or True Magistracy Restored*, 1651), which unfolded his real principles without any concealment, and propounded a complete social system based on communist principles. "The Earth," he said, "should be a Common Treasury." Particularly interesting here is his treatment of social prestige in a classless society: "As a man goes through offices he rises to titles of Honour, till he comes to the highest Nobility, to be a faithful Commonwealth man in a Parliament House. Likewise he who finds out any secret in Nature, shall have a title of Honour given him, though he be but a young

---

[1] See Berens, L. H., *The Digger Movement under the Commonwealth* (Simpkin Marshall, London, 1906); Davidson, M., *The Wisdom of Winstanley the Digger* (Henderson, London, 1904); *Gerrard Winstanley's Collected Works*, ed. Sabine (Cornell Univ. Press, Ithaca, New York, 1941). In the course of time Winstanley will come to be appreciated as standing with Milton and Bunyan as among the greatest of seventeenth-century Englishmen.

man. But no man shall have any title of Honour till he win it by industry or come to it by age or office-bearing. Every man that is above sixty years of age shall have respect as a man of Honour by all others that are younger, as is shewed hereafter."[1]

The Levellers' crisis came in 1649. In January the King was executed, in February the Council of State deliberated measures for the suppression of "disturbers of peace" in the army. Soldiers who attempted to incite the army to mutiny were to be hanged. Lilburne immediately published a pamphlet, *England's New Chains Discover'd*, against the Council of State. In March the army itself, stationed at Newmarket, protested, in a *Letter to General Fairfax and his officers*, signed by eight soldiers, who demanded the acceptance of the Levellers' *Agreement*, and who were, a few days later, after a short trial, expelled from the army. Twenty days later, the army Levellers published a pamphlet with perhaps the most remarkable title of all, *The Hunting of the Foxes from Newmarket and Thriplow Heath to Whitehall by five small beagles late of the Armie; or, The Grandee Deceivers Unmasked*. The "foxes," of course, were Cromwell, Ireton, and the rest; and their ambitious subterfuges were here exposed. A few days later there was a mutiny in London in Colonel Whalley's cavalry regiment, and, though quickly suppressed, it gave rise to a unique manifestation of popular feeling at the funeral of one of the Leveller soldiers, Robert Lockyer. I quote the account from Whitelocke's *Memorials*:[2]

"*April 29th*, 1649.

"Mr. Lockier, a trooper, who was shot to death by Sentence of the Court Martial, was buried in this manner.

"About one thousand went before the Corps, and five or six in a File; the Corps was then brought, with six Trumpets sounding a Soldier's Knell, then the Trooper's horse came clothed all over in mourning and led by a foot man.

"The Corps was adorned with Bundles of Rosemary, one-half stained in Blood, and the Sword of the deceased with them.

"Some thousands followed in Ranks and Files, all had Sea-green and Black Ribbon tied on their Hats (the Levellers' colours), and the Women brought up the Rear.

---

[1] *Works*, Sabine edn., p. 512.
[2] Whitelock, B., *Memorials of the English Affairs* (London, 1732), p. 397.

"At the new Church-yard in Westminster some thousands of the better sort met them, who thought not fit to march through the City. Many looked on this funeral as an affront to the Parliament and Army; others called them Levellers; but they took no notice of any of them."

Ten days afterwards the struggle began in earnest. News came that the troops at Banbury, Wantage, Salisbury, etc., had cast off allegiance to Cromwell, and had raised the sea-green flag in favour of the Levellers' principles. After a good deal of marching and countermarching by the Levellers and the Cromwellian praetorian troops, the former were surprised at Burford in Oxfordshire, and a fight in the streets of that town ended the chances of a second revolution. Early in June the great merchants of the City of London, who had often enough execrated Cromwell, and held tight the purse-strings in the face of the financial requirements of the parliamentary army, celebrated the overthrow of the Levellers by a splendid banquet given at Grocers' Hall in honour of Cromwell and Fairfax, the saviours of sacred property.

The Virtuosi.

Lastly, in this rapid survey of some aspects of the seventeenth century, we must give our attention to the scientific movement proceeding in quiet, apparently far removed from these excursions and alarms, but destined to be of basic social and economic importance. Out of all the events which make the seventeenth century one of the cardinal periods in the history of science, perhaps the most important was the grouping together of scientific workers of the time into societies for the furtherance of experiment and observation.[1] These societies, of which the Royal Society in England was one of the earliest, were generally under the close protection of some prince or monarch. Such royal patronage, we may believe, was dictated not so much by a purely disinterested passion for abstract truth, as by a desire for that financial prosperity which the decay of anti-usury doctrine, the urge of the rising middle class to industrial ventures, and the far-ranging thought and new technology of the scientists was combining to produce.

The Royal Society began as a group of scientists meeting both in

[1] See Ornstein, M., *The Rôle of Scientific Societies in the Seventeenth Century* (Chicago, 1928).

Oxford and in London, who called themselves the "Invisible College." The first mention of them occurs in 1646, but their incorporation under the present name did not occur till 1663. The preoccupation of the early Fellows with practical interests, with the "improvement of trade and husbandry," is patent to anyone acquainted with its early history. Thus the great Robert Boyle wrote to a friend, Marcombes:[1] "The other humane studies I apply myself to are natural philosophy, the mechanics, and husbandry, according to the principles of our new philosophical colledge, that values no knowledge, but as it hath a tendency to use. And therefore I shall make it one of my suits to you, that you would take the pains to enquire a little more thoroughly into the ways of husbandry, etc., practised in your parts; and when you intend for England, to bring along with you what good receipts or choice books of any of these subjects you can procure; which will make you extremely welcome to our invisible colledge, which I had now designed to give you a description of." Among Robert Hooke's papers in the British Museum,[2] Weld records a statement, dated 1663—"The business and design of the Royal Society is to improve the knowledge of naturall things and all usefull Arts, Manufactures, Mechanick practises, Engynes and Inventions by Experiments." Or if we look through the account and defence of the Royal Society published by Thomas Sprat, Bishop of Rochester, some years later,[3] we find that he gives a series of thirteen sample papers from the reports of the Society to show what good it has done. Of these thirteen, five are purely technical (wine, guns, salt-petre, dyeing, oysters), two are to do with exploration, and three with meteorology and astronomy, important for navigation, making a total of ten which would be "for the improvement of husbandry." The remaining three we should now call "pure science," and were devoted two to chemistry and one to physiology.

It is clear, then, that seventeenth-century science was expanding in the closest relationship with industrial enterprise. The scientific men took, indeed, little or no part in politics, but they definitely depended for their support on the party standing between and against the two groups already described (the Laudian Churchmen and the Levellers). The former were representatives of a dying pre-scientific collectivism, the latter were pioneers of a collectivism to which even yet we have

[1] Quoted by Fulton, J. F., Isis, 1932, **18**, 84.
[2] Weld, C. R., *History of the Royal Society* (Parker, London, 1848).
[3] Sprat, T., *The History of the Royal Society of London* (Knapton, London, 1722).

not attained. It was inevitable that the scientists should be with the Presbyterian and Republican "centre," the party of the rising economic individualism, since only capitalism, with its encouragement of technology, would afford science with the means for its development.

But the relation between the Presbyterian merchants and the Leveller Independents was, of course, much closer than the relations of either to the Anglican and Royalist party. Science and Industry were therefore connected broadly with the Puritan movement, a link which constantly manifests itself in this period. The complaint of a conservative Anglican divine after the Restoration is in this respect especially revealing. Samuel Parker, in his *Discourse of Ecclesiastical Politie* wrote as follows: "I confess I cannot but smile when I observe how some that would be thought wonderfully grave and solemn statesmen labour with mighty projects of setting up this and that manufacture. . . . To erect and encourage trading combinations is only to build up so many nests of faction and sedition, and to enable these giddy and humoursom people to create public disturbances. For 'tis notorious that there is not any sort of people so inclinable to seditious practises as the trading part of a nation. And if we reflect upon our late miserable distractions, 'tis easy to observe how the quarrel was chiefly hatch'd in the shops of tradesmen and cherished by the zeal of prentice-boys and city-gossips."

The Rise of Mechanistic Economics.

Drawing these many threads together now a little, we may refer to one of the most fascinating aspects of the seventeenth century, namely, the rise of "mechanistic economics." In pure science the concept of mechanical causation (or, to be more accurate, the concentration of interest on the Aristotelian efficient cause, to the exclusion of the other Aristotelian causes), was of enormous importance. No advance beyond pure descriptive biology, for instance, could be made without it. And it was just at this time that such advances were made. Thus in 1644 Sir Kenelm Digby,[1] discussing embryology, in his *Treatise on Bodies*, took the example of a germinating bean: "Take a bean," he wrote, "or any other seed, and put it in the earth; can it then choose but that the bean must swell? The bean swelling, can it choose but break the skin? The skin broken, can it choose (by reason of

---

[1] Digby, Sir K., *Two Treatises, in the one of which the Nature of Bodies, in the other the nature of Man's Soule, is look'd into, in way of discovery of the Immortality of Reasonable Soules* (Williams, London, 1664).

the heat that is in it) but push out more matter, and do that action which we call germinating? Can these germs choose but pierce the earth in small strings, as they are able to make their way? Thus by drawing the thrid carefully through your fingers and staying at every knot to see how it is tyed, you see that this difficult progresse of the generation of living creatures is obvious enough to be comprehended and the steps of it set down; if one would but take the paines and afford the time that is necessary to note diligently all the circumstances in every change of it." This was almost the first declaration of belief in the comprehensibility of the mechanism of generation. It was fundamental for the future of biology. But side by side with this, there went a similar application of mechanical causation to economics, equally fundamental for the future, but not of such happy augury. In 1622 Gervase Malynes, in his *Lex Mercatoria*, wrote:[1] "We see how one thing driveth or enforceth another, like as in a clock where there are many wheels, the first wheel being stirred driveth the next and that the third and so forth, till the last that moveth the instrument striketh the clock; or like as in a press going in a strait, where the foremost is driven by him that is next to him, and the next by him that followeth him." So men were to be thought of as selfish monads or corpuscles, like the atoms of natural science, with the automatic price-mechanisms taking the place of the Newtonian laws of motion.

Thus the theocratically legislative state of the mediaeval clerk was dying, and could no longer attempt to control the great merchants of London, Antwerp, or Venice, who looked after themselves, and expected others to do the same. Thenceforward, there was to be no interference with the free play of capitalist interests. All that was lacking was the supreme piece of cant elaborated by the eighteenth century, the opinion that when the "natural economic appetites" and selfishnesses of men are allowed to take their course to the end, a society results which does, by a strange but beneficent dispensation of Providence, provide the maximum of attainable happiness for all classes.

Mechanical causation was one concept taken over by seventeenth-century economics from science; atomism was another. It is questionable whether the social implications of these changes have been suffi-

---

[1] Malynes, G., *Consuetudo, vel Lex Mercatoria, or the Antient Law-Merchan* (London, 1622). Malynes defends usury, and describes (p. 263) certain silver mines which the owner will not allow to be worked. "Howsoever I thought good to remember this for our posteritie, for there may come a time when industrious men shall be more regarded."

ciently realised. Does not the shift of emphasis from the mediaeval state, carrying out in practice the detailed instructions of theology, *Regina Scientiarum*, to the seventeenth-century capitalists, men like Sir Thomas Gresham, pressing for the removal of every inhibiting influence on financial transactions—does not this shift of emphasis mirror in the economic sphere the transition from the four elements of the aristotelians and the three primary substances of the alchemists to the "*corpuscularian or mechanical philosophy*" of Gassendi and Boyle? "No one can deny," wrote the foreign merchants of Antwerp, about 1590, to Philip II, in a protest against an attempt to interfere with the liberty of exchange transactions, "that the cause of the prosperity of this city is the freedom granted to those who trade here." And the unrestricted competitiveness of later capitalism, the continual demand that the activities of the State should be restricted to the bare minimum necessary to safeguard property, has something so obviously atomistic about it that the nineteenth century seems surprisingly late for the appearance of the codifier of chemical atomism, John Dalton. To-day we are living in the time of the dissolution of this atomic form of society.

Of course, anarchic social atomism still (1941) finds theoretical support. M. Polanyi[1] seems to take the view that the laws of the "fortuitous concourse of particles" have the status of laws eternally applicable to human society. He first shows without difficulty that "order spontaneously arising from the mutual interactions of particles (dynamic order)," as in the crystallisation of several solutes simultaneously present in a system, much exceeds, in the inorganic world, anything that "mechanically imposed order (corporate order)" could perform. He then from this suddenly *assumes* the structure of competitive atomistic capitalism as the analogue of this dynamic order at the higher social level, perfect, ideal, and unalterable; apparently quite unconscious that the problem is what precisely *are* the mutual interactive forces between human "particles." Those who consider that anarchic social atomism has no longer a progressive role to play in human evolution do so precisely because they see that there are important interactive forces the action of which is inhibited by this system.

The fundamental drawback of atomistic capitalism[2] is that its parts

[1] In Economica, 1941, 8, 428, an article full of the most interesting fallacies.
[2] Long after making this correlation between atomism and capitalism, I found that it had been described by others also, e.g. A. M. Deborin in the Marx Memorial Volume

have no organic connection with the whole. The only concern of the industrial monads is, not that cost and price should balance but that the latter should outweigh the former by as much as they can make it. Hence a condition of universal war.[1]

It is interesting that the deficiencies of an atomistic society, as well as the tremendous changes which applied science was to bring, were realised, if dimly, by some contemporaries. Thus in 1641 Samuel Hartlib published his Utopian *Macaria*, in which it was laid down that the State alone should control and manage production as an economic institution.[2]

In 1659 Cornelius Plockboy came forward with his proposal[3] for a co-operative society, and described all the advantages which would accrue from the combination of agriculture with industry. "Whereas the Traders in the World do oppress their workmen, with heavy labour and small wages, instead thereof with us, the gain of the Tradesmen will redound to the benefit and refreshment of the Workmen."

A rather similar co-operative association was later proposed by John Bellers in 1695.[4] No practical outcome of these schemes is recorded. But the ideas of Bellers influenced Robert Owen and the Chartists two centuries later.

A new Transition.

It is my profound conviction that we are standing to-day at a turning-point between two civilisations, one of those turning-points

of the Moscow Academy of Sciences, 1933, English translation, *Marxism and Modern Thought*, p. 108. Cf. also Auden's reference to the capitalist centuries:

> On sterile acres governed by
> Wage's abstract prudent tie
> The hard self-conscious particles
> Collide, divide, like numerals
> In knock-down drag-out laissez-faire
> And build no order anywhere."
> *New Year Letter*, 1941, p. 34.

[1] See B. Wootton, *Plan or No Plan* (London, 1934), p. 159.

[2] Hartlib, S., *A description of the famous Kingdom of Macaria; shewing its excellent Government, wherein the inhabitants live in great Prosperity, Health and Happiness: the King obeyed, the Nobles honoured, and all good men respected. An example to other Nations. In a dialogue between a Scholar and a Traveller* (London, 1641).

[3] Plockboy, P. C., *A Way propounded to make the poor in these and other nations happy by bringing together a fit, suitable and well qualified people unto one Household government or little Commonwealth, etc.* (London, 1659).

[4] Bellers, J., *Proposals for raising a College of Industry of all useful Trades and Husbandry, etc.* (London, 1695).

in history not unlike the first or second christian century, the Renaissance, or the seventeenth century in England. The transition from an individualist to a collectivist state of society is at hand. In scientific words, the time has come for the atomistic, inorganic, chaotic community to give way to the organised, living, planned, community.

> "hora novissima, tempora pessima sunt, vigilemus
> ecce minaciter, imminet arbiter, ille supremus."

The times are very evil, the judge is at the gate;[1] it is the duty of the christian to join his forces with all who are seeking to bring in the new world order, the Kingdom on earth, *Regnum Dei*. As for the scientific worker, he can acquiesce no longer in the frustration of science, and must work with the rest for the overthrow of the capitalist system.

This change is hardly more likely to be achieved without tumult and civil commotion than it was likely that the middle class could peacefully overthrow the paternal-feudal system existing before 1600.

> "Comrades, my tongue can speak
> No comfortable words,
> Calls to a forlorn hope,
> Gives work and not rewards.
>
> O keep the sickle sharp
> And follow still the plough;
> Others may reap, though some
> See not the winter through."
>
> (C. Day Lewis.)

But the harshness of the days that lie before us is somewhat mitigated for the reflective mind by a clear picture of the course that history has taken. These troubles did not begin in our time; others before us have perished that the Kingdom might come. In the foregoing series of short pictures of seventeenth-century England, I have tried to show some of the forces at work.

First, the Laudian Churchmen, quite apart from the literary aspects of their brilliant scholarship and writings, were economically representative of the collectivism of the past. In the *Preces Privatae* of

---

[1] From the *De Contemptu Mundi* of Bernard of Cluny, 1145 (ed. Hoskier, H. C., London, 1929).

Lancelot Andrewes, Bishop of Winchester in 1618, there are the following words, where he prays in the manner of the Orthodox litanies, for the people of England, that they may be "subject unto rule, not only for wrath, but also for conscience' sake; to husbandmen and graziers, good seasons; to the fleet and fishermen, fair weather; to mechanics, to work lawfully at their occupations; to tradesmen, *not to over-reach one another*." And in another place, where he is rehearsing the attributes of God, he writes under the heading "Munificent"; "Opening the eyes of the blind, clothing the naked, upholding such as fall, gathering together the outcasts, giving food to the hungry, bringing down the haughty, delivering the captives, loosing the prisoners, lifting up those that are down, healing the sick, sustaining the living, quickening the dead, lifting up the lowly, helping in time of trouble." Does not this catalogue of the Divine actions curiously resemble the socialist programme? In the person of Lancelot Andrewes there is a link between the theocratic collectivism of the past and the proletarian socialism of the future.[1]

Secondly, it is in the seventeenth century that we can study the beginnings of the great scientific movement, destined so to transform the nature of civilised life in subsequent years. It was in the nature of the case that science was associated with, and patronised by, the rising middle class. And from contemporary scientific theory, indispensable for progress within science itself, bourgeois economics took its canons. But inner contradictions, given time, always come to the surface. The individualism of capitalist production was congruent enough with science in its early days, but in our time scientific effort needs a co-operative atmosphere which capitalism cannot provide. Conversely, although the bourgeois class raised itself to power partly by means of science, its need for science is now less, and its primary subconscious wish is to stabilise the existing condition of affairs. Fascism and militarism are the result.

Thirdly, the Levellers, the extreme left wing of the revolutionary forces, were really envisaging the classless State. They were far too weak, however, to make the jump across three centuries of bourgeois domination, for the economic conditions were not propitious. But is it not of some value to English socialists, tired of hearing communism

---

[1] In this connection it is interesting to find the common features in the economic systems of Aquinas and Marx emphasised by a German catholic writer, William Hohoff (*Die Bedeutung d. marxistischen Kapitalkritik*, Paderborn, 1908; and *Warenwert und Kapitalprofit*, Paderborn, 1902).

identified with foreign-sounding names and doctrines, to know that the communists of the seventeenth century had names that run like English villages—John Lilburne, William Walwyn, Gerrard Winstanley, Robert Lockyer, Giles Calvert, Anthony Sedley? So it will be again, and not for failure.

Pure Science and the Idea of the Holy

(An address delivered to the annual conference of the National Union of Students, 1941)

It is quite natural that when we stand at the beginning of a lifetime of scientific practice, whether in teaching or in industrial or "pure" research, we should feel how vital a problem is the relation between applied science and pure science. The question, moreover, is always arising, how pure should science be? What is meant by its "ethical neutrality?" Is it right or wrong for scientists to concern themselves with the social applications of their discoveries? A great deal has been written and spoken concerning these things in recent years; all I can contribute is an approach to the subject which differs somewhat from the usual approaches because of its historical and, dare I say, theological method.

Under enlightened editorship, the columns of "Nature," universally acknowledged as the world's greatest scientific weekly journal, have contained for many years past exchanges of views, sometimes put sharply enough, on the social function of science. We cannot do better than glance at two of these exchanges.

Meddling with Morals and Politics.

In 1933 the physiologist A. V. Hill discussed the general results of the scientific method during the past three hundred years in western Europe, commenting on the toleration which society as a whole has exercised towards the labours of scientific workers.[1] Scientific expeditions have been regarded as immune from the hazards of war. During the Napoleonic wars there were interchanges between French and British scientists. Science has been recognised as the common interest of mankind. "If scientific people," he went on, "are to be accorded the privileges of immunity and tolerance by civilised societies, they must observe the rules." These rules seemed to Hill to be most clearly embodied in one of the early descriptions of the aims of the newly-founded Royal Society, written in 1663, probably by Robert Hooke. "The business and design of the Royal Society is—

[1] Nature, 1933, **132**, 952.

To improve the knowledge of naturall things, and all usefull Arts, Manufactures, Mechanick practises, Engynes and Inventions by Experiments—not meddling with Divinity, Metaphysick, Morals, Politicks, Grammar, Rhetorick or Logick." "Not meddling with morals or politics," continued A. V. Hill, "such, I would urge, is the normal condition of tolerance and immunity for scientific pursuits in a civilised state."

We shall see later what was the historical context out of which this phrase was taken. Hill's point of view, expressed as it was, seemed to many to be designed to isolate the scientific worker from the outside world as much as possible. No matter what state of dire need that world might be in, of wars and oppressive tyrannies, of poverty side by side with the possibilities of utmost well-being, of widespread mismanagement and wholesale refusal to apply the lessons of scientific discovery for the benefit of mankind—the scientist should close his eyes to it, and continue his "pursuits" in courteous civility to the powers that be, not too closely examining their papers of ordination. The very word "pursuits" recalls the ideal, now somewhat threadbare, of the scientist as country gentleman, in the manner of Charles Darwin, not bound to earn any living, and existing on means not clearly specified. Hill's pronouncement quickly called forth a reply from the biochemist and geneticist, J. B. S. Haldane,[1] who had no difficulty in showing that we must distinguish between the scientist as scientist and the scientist as citizen. In his technical work, ethical and political considerations are no doubt irrelevant, but as a citizen he has a special responsibility to work for the beneficent utilisation of the discoveries of himself and his colleagues. Haldane could point to many of the most distinguished Fellows of the Royal Society who had taken this attitude, to Pepys and Brouncker, to Priestley and Franklin, of whom it was said (and could there be a more magnificent epitaph?):

"Eripuit caelo fulmen sceptrumque tyrannis."

He tore the lightning from the heavens, and the sceptre from the hands of tyrants. J. B. S. Haldane concluded, "I do not see why a man of science who 'meddles' with such matters should thereby forfeit a right to tolerance, and question whether Professor Hill has done a service to science in penning a sentence which might be inter-

[1] Nature, 1934, 133, 65.

preted as meaning that his profession should be tolerated only in so far as it is muzzled."[1]

Some seven years later, A. V. Hill abandoned his earlier position by entering the House of Commons as parliamentary representative for the University of Cambridge. The fact that on this occasion he stood as (independent) Conservative candidate, threw, in the opinion of some, an interesting light on the significance of his former entreaties to scientists to keep out of politics, but in view of the very prominent part he has taken in some of the most progressive causes of our time, such as the succouring of exiled men of science from the Continent of Europe, freely allow that he has altered his original views on the value of the public services which a scientist can render.

The Holiness of Pure Science.

A similar controversy worthy of our notice arose again in 1941. An editorial in "Nature"[2] urged that "we should abandon once and for all the belief that science is set apart from all other social interests as if it possessed a peculiar holiness." The distinguished physical chemist, M. Polanyi, rushed to the attack. "I, for one," he said,[3] "can recognise nothing more holy than scientific truth, and consider it a danger to science and to humanity if the pursuit of pure science, regardless of society, is denied by a representative organ of science. For the last ten years we have been presented by an influential school of thought with phrases about the desirability of a social control of science, accompanied by attacks on the alleged snobbishness and irresponsibility of scientific detachment. The 'social control of science' has proved a meaningless phrase. Science exists only to that extent to which the search for truth is not socially controlled. And therein lies the purpose of scientific detachment. It is of the same character as the independence of the witness, the jury and the judge; of the political speaker and voter; of the writer and teacher and their public; it forms part of the liberties for which every man with an idea of truth, and every man with a pride in the dignity of his soul has fought since the beginning

---

[1] Much the same ground as had been traversed in the Hill-Haldane controversy was gone over again by the eminent pharmacologist, Sir Henry Dale, in a Presidential Address to the Royal Society (Nature, 1941, **148**, 678; Proc. Roy. Soc., A 1942, **179**, 254; B 1942, **130**, 248), and the reply to him by the geneticist, C. D. Darlington (New Statesman, 1941, p. 524). A failure to appreciate the social responsibility of the scientist as citizen vitiates the otherwise interesting book of J. R. Baker, *The Scientific Life* (London, 1942).

[2] Nature, 1940, **146**, 815.   [3] Nature, 1941, **147**, 119.

of society." Polanyi has elaborated these views in other writings, including a vigorous attack on Soviet Science,[1] in which he takes issue with the numerous pleas advanced by J. D. Bernal and Lancelot Hogben in their books for a planned society and a socialised science and medicine.

Counsel for the defence was the experimental morphologist, C. H. Waddington. He suggested[2] that the word "holy" needed closer examination, that its use in the editorial had been to signify "esoteric," while its use by Polanyi signified "having overwhelming ethical value." No sensible person would deny the tremendous ethical value of the scientific method, or indeed of any of the other basic forms of human experience. But that is not the question. The question concerns rather its social relevance. Only on account of its social relevance does the problem of social control arise. Polanyi spoke of the independence of witness, jury, and judge. "But the witness," commented Waddington, "the jury, and the judge, turn their attention to problems presented to them as being socially important; they are not at liberty to spend the afternoon discussing the sexual habits of Polynesian worms, or whatever else takes their fancy. The editors of Nature were, as I understood them, inviting us to spend more time investigating subjects as banal but as relevant as crimes. Doubtless it is not altogether easy to preserve scientific detachment in such matters; and one can expect that what are generically termed ' powerful interests ' will attempt to influence scientific statements on matters of social consequence. But Professor Polanyi's pessimistic assumption that such influences must always be successful is vitiated by his own example of the persistence of a real and active legal detachment through many centuries of close contact with the turbulent forces of history. It may be urged that the law, though employing the methods of impartiality, is in its content merely the embodiment of the interests of the most powerful social group, and in that most important respect unfree; and it can be argued that a socially directed science, even though free to be critical and objective, would have its attention fixed down to problems chosen for it by social forces outside its control. But, speaking as an embryologist of

[1] "Rights and Duties of Science" in Manchester School of Economic and Social Studies, October, 1939, p. 175. Here he joined hands with A. V. Hill, who made a similar attack, on the occasion of his parliamentary candidature; for this and for my reply, see New Statesman, 1940, pp. 105, 174, 206. See also M. Polanyi's book, *The Contempt of Freedom* (London, 1940).

[2] Nature, 1941, **147**, 206.

no cash value to anybody, and addressing a physical chemist of great industrial importance, I should like to ask if something of this sort is not true already? Our civilisation is, to some degree, a society, and not a mere collection of individuals. Men of science are, again to some degree only, involved in the social bonds which create society's coherence. One could only be justified in calling for a less degree of involvement in those bonds if one disapproved of the society as a whole, if, for example, one was a revolutionary who wished to stay outside it so as to overthrow it." And no one could suppose that this description could apply to Prof. Polanyi.

A society and not a mere collection of individuals. Has not science itself risen to its present position of domination over nature precisely by virtue of taking social coherence seriously? Is not modern science distinguished from primitive forms of science such as alchemy by the fact that the free publication of results permits of confirmation, or failure of confirmation, by a thousand observers and experimentalists, scattered over the earth's surface, of every race, religion, colour and creed? Does not science strive to perfect the means for the communicability of human thought about nature? Science is a society within a society. Its social nature and function are inescapable. Its holiness, or the reverse, cannot be thought of without considering the holiness of society as a whole. We shall return later to this question.

Science, Authority and Freedom.

Here I would interject an extremely important matter to which C. H. Waddington, in another place,[1] has recently drawn attention. Science, he says, is the most perfect resolution which man has yet found, of the antinomy between Authority and Freedom. The structure of human scientific knowledge about the world is never complete; there is always the possibility that some fundamental discovery may be made which will require the modification of at any rate large parts of it. Such was the case with the theories of Einstein on space and time. Authority in science can never, therefore, be absolutely secure. It is open to anyone to upset the whole structure or a large part of it, *if he can*. And it is part of the spirit of science that he should try his best to do so. *Freedom* is therefore secure, but in so far as the greater part of scientific knowledge is solidly established, and therefore has a certain *Authority*, Freedom becomes indeed the

[1] *The Scientific Attitude* (Penguin Books, London, 1941), p. 93.

Knowledge of Necessity.[1] These considerations have an important bearing on the concept of the ideal human society. It ought to be so rationally constructed, based so soundly on the ascertained nature and needs of human beings, that though it would be open to anyone to question the foundations of it, or of a large part of it, the chances of the substantiation of a need for a radical reconstruction of it would be extremely small. Since it is only by the scientific method that such a society could be formed in the first place, we see a further significance in the phrase "scientific socialism" adopted by the earliest exponents of marxist theory. The phrase was used because this form of socialism was based, not on utopian hopes, but on a conviction of the continuity of social with biological evolution, and hence a conviction of the inevitability of higher forms of social organisation than those we now possess. But it acquires a further significance when we realise that the society of the future must be one so founded on reason that its rulers can afford the luxury, hitherto unattained by any rulers, of being open to conviction. A rational social system, as the world co-operative commonwealth would be, would have nothing to fear from the upwelling of its own irrational contradictions, but would be open, just as the system of science is to-day, to proposals for change. Let any comer better the system, *if he can.*

Let us now return to the historical origins of science in western Europe, and particularly to the seventeenth century, the age of the foundation of the Royal Society, from whose archives A. V. Hill fetched his useful phrase.

The political background of the early Royal Society.

Modern historical research has established beyond question that the great movement of science which, though extending back into the fifteenth century, achieved its most magnificent victories and attained its maximum rate of progress in the seventeenth, was one of a number of vast social processes all taking place at about the same time. The transition from the period of mediaeval stagnation in science paralleled the economic transition from feudal, social and economic forms to those of capitalism, and this involved a general shift of emphasis from agricultural

[1] Is it not interesting that we find in early Chinese thought a premonition of this? In the Kuan-tze, a philosophical work of about 100 B.C. (ch. 18) we find: "The sage follows after things, in order that he may control them." (Shêng-rjêng ying chih, ku nang ch'ang chih.)

production in small closed communities to industrial production in large open ones. The country declined in importance and the cities rose to power. The dominance of an aristocracy based on land ownership gave place to the dominance of a merchant society based on the possession of monetary wealth and the power to employ it in profitable enterprises. To these changes in the world of daily life there corresponded changes in the world of ideas, of philosophy and theology, and here the transition was represented by the Reformation and all that that implied. Merit, formerly acquired by the contemplative life, was now to be gained by the active life of service (not without reasonable material recompense) to one's fellow-men. Wealth ceased to be a sign of diabolic favour and became a sign rather of the favour of God. The protestant and puritan movement was revolutionary not only in theology, but in public life too. The right of christian people to rebel against unchristian tyrants was invoked in the cause of protestantism, as by Bishop John Ponnet.[1] The soldier-preachers of the Parliament's Army in the civil war period began by finding no warrant in scripture for bishops or presbyters; they went on to find no warrant there for landlords either. The mediaeval restrictions on usury stood in the way of the new economic progress; if some of the finest scholars and writers of Caroline England, such as Jeremy Taylor and Lancelot Andrewes, stood in the way, so much the worse for them.

It is almost unnecessary to ask on what side stood the protagonists of that young giant awaking from his sleep, the scientific movement itself. They were almost to a man associated with the progressive social trend of protestantism. Miall, in his classical book on the lives of the early naturalists, has shown the great preponderance of protestants among the botanists and zoologists of the late sixteenth and early seventeenth centuries.[2] Only one of the men who came together in the year 1649 to discuss scientific subjects and make co-operative scientific experiments, in what was then known as the "Invisible College" and afterwards became the Royal Society, was of royalist affiliations, the physician, Charles Scarborough. And he was personally connected with the only scientist of first-class importance in the English seventeenth century, who was more or less on the royalist side, namely the great William Harvey, physician to King Charles I.

[1] *A Short Treatise of Politique Power* (1556), the first book the Parliament reprinted in their propaganda campaign of 1642.
[2] L. C. Miall, *The Early Naturalists; their lives and work* (London, 1912).

In an important monograph, R. K. Merton has conclusively shown that the affiliations of all the early Royal Society Fellows, with very few exceptions, were with the protestant and parliamentary side.[1]

In the light of facts such as these, the statement quoted by A. V. Hill is to be seen in a rather different light. The early Fellows of the Royal Society were no doubt, as a body, desirous of taking no particular political line, but that does not mean that individually they had no political sympathies. There is significance in their meeting-places, first at Oxford, then at Gresham College in London, and later their close connection with Cambridge, the home of their greatest ornament, Isaac Newton. The University of Oxford, where they met first, and where a marked group of scientific men had gathered, was not definitely royalist, as it became during the civil war. Posts there were later taken from their holders and given to royalist supporters; thus William Harvey was made Warden of Merton, but wisely rode out over Shotover Hill with the retreating royalist army when Oxford finally fell to the Parliament's arms. In London, Gresham College had been founded by one of the most famous of London's sixteenth century financiers, Sir Thomas Gresham, a member of my own Cambridge College, Caius. He established there professors of such useful sciences as astronomy, geography, navigation, and the like, with the avowed intention of training the technicians required by the new expanding capitalist enterprises. His school at Holt in Norfolk could take care of less advanced education. And his third foundation, the Royal Exchange, had a connection with the power of the City which needs no emphasis. That the movement of science passed from Oxford before the civil war to Cambridge afterwards is also interesting, for Cambridge had been almost wholly on the progressive, revolutionary, puritan, parliamentary side. Only two colleges sent plate to the King's treasurer, and that had been neatly recovered before it ever reached him. Cambridge was the headquarters of the Eastern Counties Association, the backbone of the parliamentary army before the establishment of the New Model. It was held for Parliament throughout the war.[2] What manner of men accepted posts in the University under the revolutionary auspices may be seen by such examples as Benjamin Whichcote and William Dell.

[1] R. K. Merton, "Science, Technology and Society in seventeenth century England" in Osiris, 1938, 4, 360.

[2] Cf. *Cambridge during the Civil War* by F. J. Varley (Cambridge, 1935), and *East Anglia and the Civil War* by A. Kingston (London, 1897).

Benjamin Whichcote, a gentle Neo-Platonist, made Provost of King's, proceeded to divide his salary in half and share it with the extruded royalist. William Dell, made Master of Caius, used his opportunities for urging a vast extension of educational facilities, at that time unheard of. He believed that the Universities should banish the classics, and teach rather "logic and mathematics, but especially arithmetic, geometry, geography, etc., which as they carry no wickedness in them, so are they besides very useful to human society." He also wrote against the monopoly of learning by Cambridge and Oxford, suggesting that it would be "more advantageous to the good of all the people, to have universities or colleges, one at least, in every great town or city in the nation, as in London, York, Bristol, Exeter, Norwich, and the like; and for the State to allow these Colleges an honest and competent maintenance, for some godly and learned men to teach the tongues and arts there, under a due reformation." After the Restoration, Dell retired to the country, and did not escape the following accusations:—

"He has reported that the King and his followers were like the Devil and his angels, and has approved the murder of the late King, and the taking away of the House of Lords. He has entrapped the gentry of the county into discourses and then given information against them. He has declared in the public congregation that he had rather hear a plain country man speak in the church, that came from the plough, than the best orthodox minister that was in the country; upon Christmas Day last, one Bunyan, a tinker, was countenanced and suffered to speak in his pulpit to the congregation, and no orthodox minister did officiate in the church that day. Since the restoration of the expelled members of parliament he has declared that the power was now in the hands of the wicked, and that the land was like to be flowed over again with popery. He has put forth several seditious books, and before the horrid murder of the late King he declared publicly to his congregation that the King was no king to him; Christ was his King; Venice and Holland were without earthly kings; why might not we be without? and that he did not approve of earthly kings."

Not bad for a Master of a College.

Science and Society in the English Revolution.

But to return to the Royal Society. That the first meetings of it took place about 1649 is a remarkable historical coincidence. For that was the year which, after the execution of the King in January, saw during the summer the lightning campaign in which Cromwell and Ireton, with their praetorian guard of dependable troops, put down, at the battle of Burford, the armed attempt of the Parliamentary Army's left wing, the Levellers, to attain power in order to implement the *Agreement of the People*. Here we cannot tell the story of those remarkable days.[1] The *Agreement*, though not what we should regard to-day as a socialist document, was a great deal beyond anything that the eighteenth or even the nineteenth centuries achieved.

The Levellers comprised, besides their military wing, at one time dominant over all other feeling in the army, a civilian, pacifist, wing, known as the Diggers. Gerrard Winstanley, the leader of this group, which mainly devoted its efforts to the initiation of co-operative farming, a venture which did not succeed owing to opposition both local and central, was clearly aware of the importance of science, the "new or experimental philosophy," for social welfare. This is shown by his books *A New Year's Gift to the Parliament and the Army*; *The Law of Freedom in a Platform*, and *The True Leveller's Standard Advanced*, all published about the year 1650.[2]

Winstanley discusses science in relation to religious services, which in those days were a medium for the dissemination of news and popular education as well as for public worship.

> "If the earth were set free," he said, "from kingly bondage, so that every one might be sure of a free livelihood, and if this liberty were granted, then many secrets of God and his works in Nature would be made public, which men do nowadays keep secret to get a living by; so that this kingly bondage is the cause of the spreading of ignorance in the earth. But when the Commonwealth's freedom is established, then will knowledge cover the earth as the waters cover the seas, and not until then. He who is chosen minister for the year shall not be the only man to make sermons or speeches (on the day of rest from labour);

[1] See *The Levellers and the English Revolution* by Henry Holorenshaw (London, 1939).
[2] The collected writings of this great Englishman have just been published by Prof. G. Sabine in the U.S.A., and a selected edition will, it is hoped, appear in this country under the editorship of Mr. L. Hamilton.

but everyone who has any experience and is able to speak of any art or language, or of the nature of the heavens above or of the earth below, shall have free liberty to speak when they offer themselves and in a civil manner desire an audience; yet he who is the reader for the year may have his liberty to speak too, but not to assume all power to himself, as the proud and ignorant clergy have done, who have bewitched all the world by their subtle covetousness and pride. And every one who speaks of any herb, plant, art, or nature of mankind, is required to speak nothing by imagination, but what he has found out by his own industry and observation in trials [experiments]. And thus to speak, or thus to read, the law of Nature (or God) as He hath written His name in every body, is to speak the truth as Jesus Christ spake it, giving to everything its own weight and measure. 'Aye, but' saith the zealous but ignorant professor [of religion], 'this is a low and carnal ministry indeed. This leads men to know nothing but the knowledge of the earth, and the secrets of Nature; but *we* are to look after spiritual and heavenly things.' I answer, to know the secrets of Nature is to know the works of God, and to know the works of God within the creation is to know God himself, for God dwells in every visible work or body. And indeed, if you would know spiritual things, it is to know how the Spirit or Power of Wisdom and Life, causing motion or growth, dwells within and governs both the several bodies of the stars and planets of the heavens above, and the several bodies of the earth below; as grass, plants, fishes, beasts, birds, and mankind; for to know God beyond the creation, or to know what he will do to a man after the man is dead, if any otherwise than to scatter him into his essences of fire, water, earth, and air, of which he is compounded; is a knowledge beyond the line or capacity of man to attain to while he yet lives in his compounded body."[1]

This wonderful passage demonstrates that even among a true representative of the people (for Winstanley was not of middle-class origin, like most of the early scientists) a clear understanding of the social importance of science could be found. Both Winstanley and the early Fellows of the Royal Society, good puritans, gave as their conscious motive the study of the works of God.

[1] *Works*, Sabine edn. (Cornell, 1941), p. 564.

I shall give two other instances of the close connection between revolutionary politics in the seventeenth century and strongly progressive ideas in science. Puritans coupled intense scorn for "jejune Peripatetick Philosophy" (i.e. the outworn Aristotelianism) with admiration for "mechanical knowledge." Noah Biggs, a staunch puritan, wrote a book entitled *Mataeotechnia Medicinae Praxeos* (1651), dedicated to Parliament and Cromwell, and calling on its title-page for "a thorough reformation of the whole art of physick." He put his finger on the right spot in asking "Wherein do the Universities contribute to the promotion or discovery of truth? Where have we anything to do with Mechanicall Chymistrie, the handmaid of Nature, that hath outstript the other sects of philosophy, by her multiplied real experiences? Where is there an examination and consecution of Experiments? encouragements to a new world of knowledge, promoting new Inventions? where have we constant reading upon either quick or dead Anatomies? or an ocular demonstration of Herbs? Where a Review of the old Experiments and Traditions and a casting out of the rubbish that has pestered the Temple of Knowledge?"

So also that remarkable man, Marchamont Needham. From 1643 to 1647 he edited the newspaper of Parliament, in opposition to the royalist newspaper edited by Sir Roger l'Estrange. He then started a new paper, more royalist in tone, for which he has often been condemned by subsequent writers as a turncoat, but when the facts are closely examined, it is seen that this was just the time when Cromwell and indeed all parties were trying to get the King to come to some compromise. The change was thus more apparent than real, as is further shown by the violence with which the new paper attacked the Presbyterian Scots, even before they had invaded England to restore the King. After the King's death and the establishment of Cromwell as Protector, Needham threw off conciliatory masks and published *The Case of the Commonwealth of England Stated*, in which he showed himself almost the only writer then living who fully understood that he was in a revolutionary period. For the rest of the Commonwealth period he continued to edit the Parliament's newspaper, in co-operation with John Milton. But now, at the Restoration, after he had retired from public life, and taken up again the practice of medicine, he produced his *Medulla Medicinae* (1665). In this book he made an intensive attack on the old-fashioned reliance on Galenic or herbal remedies in therapy, thus aligning himself with the followers of Paracelsus, such as Thomas Willis and Thomas Dover, who did in the end succeed in introducing into the pharmacopoeia such valuable drugs as antimony, mercury, alum, bismuth, and other "chymical" remedies. Even in this latter part of his life Needham was a storm-centre; of the drug controversy he wrote, "Four champions were employed by the Colledge of Physicians to write against this book; two died shortly afterwards, the third took to drink, and the fourth asked my pardon publicly, confessing that he was set on by the brotherhood of that confederacy!"

What of the Restoration? It is often misunderstood. Though accompanied by repressive measures, no doubt, against the William Dells and the Needhams and Winstanleys, it was a restoration, not of the absolute Stuart monarchy, but an instauration of something very different, something that was soon to become, in so far as it was not already, a constitutional government in which the King would govern, if he governed at all, by the grace of the bourgeoisie embodied in the power of the City of London. It was, in fact, a compromise, or perhaps a dialectical synthesis arising out of the former deadlock. When Monk called on old retired Fairfax at his home in Yorkshire in 1660, "My Lord told him his mind, that there was no peace or settlement to be expected in England, but by a *free* Parliament *and* upon the old foundation of Monarchy." Though the Parliament's dead may have turned in their graves to see a King at Whitehall again, they need have had little anxiety. It was not what they had expected, but it was quite different from what they had destroyed. The English revolution had done what all successful social revolutions do, it had gone two steps forward and only one step back. The bourgeoisie was now in power. Hence the royal patronage extended to the scientists in their labours, labours which, as almost anyone could see, would be for "the improvement of trade and husbandry." It was their own phrase.

This close association of the early Royal Society with practical industrial development has puzzled many later writers, but it is now a well-known story, and has even had special books, such as that of G. N. Clark,[1] devoted to it. "The noise of mechanick implements," wrote Bishop Thomas Sprat,[2] the Royal Society's first historian, "resounds in Whitehall itself." It would not have done so in a Caroline court. So, too, about the same time, science in education,

[1] *Science and Social Welfare in the Age of Newton* (Oxford, 1937).
[2] *History of the Royal Society* (London, 1670, often reprinted, e.g. 1722).

urged by the great Comenius, became respectable. "I have often thought it a great error, wrote Bishop Burnet, "to waste young gentlemen's years so long in learning Latin by so tedious a grammar. I know those who are bred to the professions in literature, must have the Latin correctly.... But suppose a youth had, either for want of memory or application, an incurable aversion to Latin, his education is not for that to be despair'd of; there is much noble knowledge to be had in the English and French languages; geography, history, chiefly that of our own country, the knowledge of Nature, and the more practical parts of the Mathematicks... may make a gentleman very knowing, tho' he has not a word of Latin."

But long before, in Francis Bacon's writings, that great man born out of due time and languishing in an age just before the dawn, there were constant indications of the practical outlook. In *Valerius Terminus*,[1] for example: "And therefore it is not the pleasure of curiosity, nor the quiet of resolution, nor the raising of the spirit, nor the victory of wit, nor faculty of speech, nor lucre of profession, nor ambition of honour or fame, nor inablement for business, that are the true ends of knowledge; ... but it is a restitution and re-investing (in great part) of man to the sovereignty and power (for whensoever he shall be able to call the creatures by their true names he shall again command them), which he had in his first state of creation." And later, "The dignity of this end, the endowment of man's life with new commodities, appeareth by the estimation that antiquity made of such as guided thereunto. For whereas founders of states, law-givers, extirpers of tyrants, fathers of the people, were honoured but with the titles of Worthies or Demigods, inventors were ever consecrated among the Gods themselves." And again, in the *Filum Labyrinthi*,[2] "He [Bacon himself] saw plainly, that this mark, namely invention of further means to endow the condition and life of man with new powers or works, was almost never yet set up and resolved in man's intention and enquiry."

These passages, among many others which could be quoted, though indicating the traces of magical ideas which existed in the Baconian concept of foolproof inductive methods, abundantly show how central was his emphasis on the improvement of the lot of mankind which science would achieve. A lonely figure, the "Bell that called the Wits together" rang with very good effect. And the close association of the Royal Society with trade and husbandry lasted on

[1] *Works*, ed. Ellis & Spedding, p. 188.   [2] Ibid., p. 208.

into the eighteenth century, until a new economic situation and new imperial conquests brought a new equilibrium of classes in which the Royal Society turned its attention rather to "pure" science. Then the centre of gravity of applied science shifted to the north, where the "Lunar Society" at Birmingham, and certain Scottish groups, took up the tale.

Evidently, there is a great deal more in the proposal of the Fellows of 1663 not to "meddle with morals or politics" than meets the eye.

So much by way of commentary on the case of Hill v. Haldane. We may now consider the case of Polanyi v. Waddington. What exactly do we mean by pure science?

The Meaning of Pure Science.

Everyone agrees that the essence of science is the spirit of free enquiry. Science is, as Polanyi says, a body of valid ideas about Nature. The basic test of scientific truth is whether or not it fits in with the total body of scientific concepts which has grown up through centuries of human effort. The first necessity, when a new group of facts has been established by competent workers armed with the appropriate techniques, and duly confirmed, is to elaborate some hypothesis linking it up in a rational way with the existing body of scientific knowledge. The predictability of events is taken as justification for the belief that through the centuries science is approximating more and more closely to truth. Science is autonomous in that certain modes of reasoning are entirely foreign to it, and ethically neutral in that natural phenomena are unaffected by our desires. They represent what is, irrespective of what we think ought to be.

But science does not exist in a vacuum. It is essentially a product of society, and the communism of its co-operating observers is but a prefiguration of that economic and social solidarity which humanity is destined to achieve. The social background of science influences it in many ways. Among the most important of these are the factors which stimulate the interest of a scientist at a given moment in history in one direction rather than another. Out of the infinity of possible problems which he might attack, he does in fact attack only a certain problem since life is short and the art long. The history of science abounds in illustrations of this, starting from primitive examples such as the origin of alchemy in the search for the Pill of Immortality in China and the Philosopher's Stone in the West, to the factors in seventeenth-century economic life, which, as Hessen[1] showed in a classical essay, directed the attention of Newton into specific channels. Works such as those of Singer,[2] Pledge,[3] and Crowther[4] give many more such instances. Genetics, to pass to biology, did not originate out of pure rationalistic thought of Greek type, but in close association with problems of plant breeding, in the garden of an abbey. Nor are the departments of science themselves in watertight compartments; the observations of botanists gave rise to the physical chemistry of membrane permeability, and the study of monomolecular films would perhaps never have arisen without the biochemistry of lipoidal materials.

The fears expressed regarding pure science really imply that social forces external to scientists may attempt to confine their activities to "applied" problems of short-term scope. Were this to happen anywhere it would indeed be a calamity. It would be killing the goose that lays the golden eggs. The development of scientific thought, proceeding according to its own inner logic, and free to direct its attention to whatever facts may seem relevant, is the only guarantee that discoveries of fundamental importance to humanity will be made. Societies which sought to save their life by narrow scientific concentration on short-term problems, would certainly lose it. But this attitude is to be found neither in the democracies nor in that bugbear of Polanyi's, the Soviet Union. We may illustrate the Russian marxist policy in these matters by one very striking example, namely the support which has been given in Russia to the sciences of experimental morphology and embryology. These subjects are of quite fundamental importance for biological thought since they elucidate the laws of the coming-into-being of organisms, the fixation of fates, the morphogenetic hormones, the onset of differentiation by function, etc., etc. But they stand at the furthest remove from likely practical applications, either in war or peace. The two countries where they were most extensively studied, up to about 1920, were America (for invertebrates) and Germany (for vertebrates); England, unfortunately, has never had a vigorous embryological tradition. But during the past ten years these important subjects have almost ceased to exist in Germany, while Russia, building on substantial previous foundations,

[1] B. Hessen, "The Social and Economic Roots of Newton's Principia" in *Science at the Cross-roads* (Kniga, London, 1931).
[2] C. Singer, *A Short History of Science* (Oxford, 1941).
[3] H. T. Pledge, *Science since 1500* (London, 1939).
[4] J. G. Crowther, *The Social Relations of Science* (London, 1941).

has more than taken Germany's place, and America has benefited by a large number of first-rate morphologists exiled from Germany.

Again, fears have sometimes been entertained that the sciences of evolution, palaeontology and the like, would suffer in a socialist society. But in point of fact, they have a fundamentally important *use*, the construction and elaboration of that true world-view on which "scientific socialism" determined to base itself, apart from the fact that they are inextricably intertwined with matters of practical use in the ordinary sense. Thus the school of geochemistry for which Russia is famous, headed by Vernadsky,[1] helped to reveal the course of the earth's evolution as well as the sites of mineral resources. And the unrivalled collection of cultivated plant varieties from all over the world made by the school of Vavilov,[2] threw much light on the prehistoric origin and development of agriculture as well as providing unfamiliar and desirable types of plant for practical cultivation.

All this is a remarkable commentary on Polanyi's fear that "pure" science has been banished from marxist Russia.[3] It is a fear without rational basis.

There is really no distinction between "pure" and "applied" science. The common distinction between science "for its own sake," and science "for its practical usefulness" is unsound.[4] Human motivations

[1] See W. Vernadsky, *La Biosphère* (Paris, 1929).
[2] Now (1942) a Foreign Member of the Royal Society.
[3] A few further words may be added regarding the attacks of Hill and Polanyi on Soviet science. One may grant them that the language of editors of Soviet periodicals has been uninhibited and lacking in old-world courtesy. One cannot but regret the difficulties which individual scientists may have got themselves into with the Soviet government, although in the absence of the full facts, any decision on such matters is difficult. One cannot defend the lesser degree of individual liberty which is thought to have prevailed in the Soviet Union during the past twenty years, except by pointing out that the period of dictatorship of the proletariat involved methods of defence unnecessary in the older democratic countries. Such methods were employed in the belief that the Soviet Union would probably be attacked by the fascist powers, a belief now shown to have been only too well founded. Polanyi draws attention to the controversies in the Soviet Union on genetics and psycho-analysis. For the former the reader is referred to an article (Modern Quarterly, 1938, **1**, 369) which shows that far too much has been made of the genetics controversy in this country. As regards psycho-analysis, the remark said to have been made by a psycho-pathologist is worth quoting. "In Russia, people can marry whom they like; that would take away half my practice. They can also divorce whom they like. That would take away the other half." As for Professor Kapitza, who is represented by Professor Polanyi as a haggard political prisoner, he has now (1941) received the highest Soviet decorations and has spoken on the radio to British scientists as the mouthpiece of Soviet physicists.
[4] Though still maintained by eminent scientists, e.g. A. G. Tansley in the Herbert Spencer Lecture at Oxford for 1942.

are always too mixed. There is really only science with long-term promise of applications and science with short-term promise of applications. True knowledge of nature emerges from both kinds of science.

A few words may be added about the planning of scientific research. Polanyi, and Baker (in *The Scientific Life*), with their individualist bias against the modern trend towards more adequate planning, go about to alarm us of the evils of an excess of it. They would have us believe that planning is not compatible with "pure," though it may be with "applied," science. But the positive advantages of planning are agreed to upon all sides; it is essentially only an extension of broadly-conceived research programs such as already exist, it is necessitated by that ever-increasing complexity of research which demands collaborative methods, and it will facilitate the interrelations of specialised workers. The kernel of truth in the protests of Baker and Polanyi is that scientific discovery partakes of the nature of the creative arts, and that scientists can not guarantee to produce results to a timetable. Russians, however, are not often accused of lack of understanding of art and artists. Their scientific planning has indeed, for the most part, been broadly conceived. Moreover, it ill becomes the capitalist countries to criticise any growing pains of socialist planning, for the pressure on the young scientific worker to produce results at all costs has nowhere been more acute than in some American institutions, and is far from unknown in England.

Pure Science and Philosophy

Or again, it may be held that scientific research can only be pure when it is conducted in an atmosphere unpolluted by any particular philosophy. That the influence of an official state world-outlook, philosophy, or set of theories can be catastrophically bad for science is seen from the condition of science under nazism and fascism. The principle of "race-conditioning" of scientific thought has led to an infinity of absurdities in the German literature, and to a substantial fall in productivity in many branches of science, both in bulk and quality, by the exiling of many of the most distinguished German scientists. But Soviet Russia also has a philosophy accepted by the dominant party in the state, that of marxist dialectical materialism. On the value of this to science, opinions differ. Polanyi and a number of other British scientists dismiss it as nonsense, though never clearly stating their objections to it. Some of us, on the other hand, in growing number, believe it to embody a world-outlook of great value and to be capable of providing the working researcher with a reliable guide to his thought, not indeed telling him what he is likely to find, but ensuring that when he has found it he shall avoid making the various kinds of mistakes which builders of scientific theory have often made.

If we look back at the beginnings of the scientific movement in England and throughout Europe, we see at once that the "founding fathers" of science were by no means indifferent to philosophy. The rise of the "new or experimental philosophy" was carried through to the accompaniment of a furious battle with the surviving representatives of the Aristotelian or scholastic tradition. In order to gain some understanding of what the early scientific workers were up against, every student of science to-day should read the *Scepsis Scientifica* of Joseph Glanville,[1] or the classical essay on this seventeenth-century struggle by Francis Gotch.[2] In the year 1631 a young man delivered in the Old Schools of Cambridge University, only a stone's throw from where I write these lines, an "academic prolusion" in the form of a frontal attack on scholastic philosophy. That young man was John Milton, later to be Latin or Foreign Secretary of the Commonwealth of England. The studies of scholastic philosophy, he said:[3]

> "are as fruitless as they are joyless, and can add nothing whatever to true knowledge. If we set before our eyes those hordes of old men in monkish garb, the chief authors of these quibbles, how many among them have ever contributed anything to the enrichment of literature? Beyond a doubt, by their harsh and uncouth treatment they have nearly rendered hideous that philosophy which was once courteous, well-ordered, and urbane, and like evil genii they have implanted thorns and briars in men's hearts, and introduced discord into the schools, which has wondrously retarded the progress of our scholars. For these quick-change philosophasters of ours argue back and forth, one bolstering up his thesis on every side, another labouring hard

---

[1] *Scepsis Scientifica; or Confest Ignorance the Way to Science, in an essay on the Vanity of Dogmatising and Confident Opinion* by Joseph Glanville, F.R.S. (London, 1661, reprinted 1885).

[2] In *Lectures on the Method of Science*, ed. T. B. Strong (Oxford, 1906). For other details on the twilight of Aristotelianism, see my *History of Embryology* (Cambridge, 1934).

[3] John Milton's *Private Correspondence and Academic Exercises*, tr. P. B. Tillyard, ed. E. M. W. Tillyard (Cambridge, 1932). The third Prolusion, pp. 67 ff.

to cause its downfall, while what one would think firmly established by irrefragable arguments is forthwith shattered by an opponent with the greatest ease...."

In this attack Milton was thus at one with another great patron saint of the scientific era, the Czech Comenius (Jan Amos Komenský), who strove through forty years of exile in nearly every European country for an education based on *things* and *actions*, not *words* and *ideas*,[1] and who was later personally known to Milton when he visited England in 1641 at the request of Parliament to plan a reform of our educational system. But we are back in the schools at Cambridge ten years before. Milton is going on, in his sonorous Latin, to say:

"How much better were it, gentlemen, and how much more consonant with your dignity, now to let your eyes wander as it were over all the lands depicted on the map, and to behold the places trodden by the heroes of old, to range over the regions made famous by wars, by triumphs, and even by the tales of poets of renown, now to traverse the stormy Adriatic, now to climb unharmed the slopes of fiery Etna, then to spy out the customs of mankind and those states which are well-ordered; next to seek out and explore the nature of all living creatures, and after that to turn your attention to the secret virtues of stones and herbs. And do not shrink from taking your flight into the skies and gazing on the manifold shapes of the clouds, the mighty piles of snow, and the source of the dews of the morning; then inspect the coffers wherein the hail is stored and examine the arsenals of the thunderbolts. And do not let the intent of Jupiter or of Nature elude you, when a huge and fearful comet threatens to set the heavens aflame, nor let the smallest star escape you of all the myriads which are scattered and strewn between the poles; yes, even follow close upon the sun in all his journeys, and ask account of time itself and demand the reckoning of its eternal passage."

In such a sublime manner did the spokesmen of the scientific movement inaugurate their plan. What was inconsistent with its free

[1] Cf. the volume published on the occasion of the Tercentenary of Comenius' visit to England—*The Teacher of Nations*, with essays by President Beneš, J. L. Paton, J. D. Bernal, Ernest Barker and others (Cambridge, 1942).

development had to go; and the Aristotelian concepts of causation, of the four elements, of form and matter, of qualities and entities; the Galenic concepts of virtues and humours, *were* inconsistent with it. It was no chance that science was called then and for long afterwards, "natural philosophy." But we have little reason for supposing that any improvement on the philosophy which grew up with the scientific movement is now for ever impossible.

One thing at least is clear; it is impossible for a scientist to have no philosophy at all. The freer from it he thinks he is, the more surely he is in the grip of some unconscious system, perhaps a garbled form of idealism which he received as a virus in the milk of the religion on which he was nourished as a child.[1] Purity of science in this sense is an illusion. In the christian western democracies, theological idealisms of various kinds have long overlain the materialism of primitive christianity. One might almost take leave to doubt whether the philosophy of dialectical materialism is more dominating in the minds of Soviet scientists, than the mystical idealism of Sir Arthur Eddington and Sir James Jeans in the minds of British scientists who firmly believe themselves to have no philosophy at all. But if one must have a philosophy, let it be a good one, congruent with the scientist's experience of nature. The dialectical materialism of Marx and Engels, the organic mechanism of Whitehead,[2] the emergent evolutionism of Lloyd-Morgan[3] and Smuts,[4] the temporal realism of Alexander,[5] the evolutionary naturalism of Sellars,[6] and many other writers; all these

---

[1] Cf. the words of Engels:

"Scientists imagine that they can free themselves from philosophy by ignoring or disdaining it. But as they are unable to move a step without thought, and thought demands logical definitions, the only result is that they take these definitions uncritically either from the current ideas of so-called educated people, dominated by hang-overs from philosophical systems long since decayed, or else from their random and uncritical reading of all kinds of philosophical works. In fact, they prove themselves the prisoners of philosophy, but unfortunately on most occasions, of philosophy of the worst sort. Thus while they are most violent in their contempt for philosophy they become the slaves of the most vulgarised relics of the worst philosophical systems." (*Dialectics of Nature*, Gesamtausgabe edition, Moscow, 1935, p. 624.)

[2] See another essay in this book, p. 178.

[3] Gifford Lectures, *Emergent Evolution* and *Life, Mind and Spirit* (London, 1923 and 1926). It is interesting that Lenin in his *Materialism and Empirio-Criticism* (*Works*, 11, p. 244), approved of Lloyd-Morgan's earlier work, in which he attacked the Machian, Karl Pearson.

[4] *Holism and Evolution* (London, 1926).

[5] *Space, Time and Deity* (London, 1927).

[6] *Evolutionary Naturalism* (Chicago, 1922).

have much help to give to the scientific worker and the greatest of them is the first.[1]

Political Distortions of Scientific Ideas.

There is one sense, however, in which the scientific worker should be perpetually on his guard as to the purity of science. Every form of human activity which has elaborated a theoretical structure has had this theoretical structure systematically distorted as part of the exploitation of all ideas in the interests of dominant classes. The history of theology, for example, shows the emphasis laid on pietistic and other-worldly conceptions, an emphasis much to the advantage of those who were doing very well out of this world and were determined to do better. Hence the condemnation of religion as the opium of the people—not a new idea at all, but clearly expressed by Gerrard Winstanley and Richard Overton[2] in seventeenth-century England. So also historians of the Roman religion of pre-christian times, such as B. Farrington,[3] have elucidated the class elements in the opposition of the Epicureans to the superstitious beliefs and rites cynically imposed by governments of wealthy sceptics. This is indeed quite explicit in Cicero's *De Natura Deorum*.[4] Was not the doctrine of original sin also of use to the governing class? Did it not, by suggesting a fundamental hopelessness regarding the improvement of human society, insinuate that the masses would never learn to rule themselves?

So it has been with ideas of a scientific order. Among the oldest of these is the comparison between the social organism and a living biological organism. The social organism was supposed to have also its brain, its belly and its legs. And just as the upper parts of the body, the brain, the organs in the thorax, etc., were supposed to be nobler and more honourable than those in the viscera or limbs, so obviously, it was argued, there must always be some classes of society possessing honours and privileges, while there must always be others deprived of them. The attribution of ethical superiority to some of the organs of the body is so strange an idea in itself that it must surely be regarded

---

[1] Perhaps the best discussions of the help given to biological thinking by dialectical materialism are in the contributions of H. J. Muller and J. Schaxel to the Lenin Memorial Volume of the Moscow Academy of Science (1934).
[2] *Man's Mortalitie*, 1643.
[3] Modern Quarterly, 1938, **1**, 214.
[4] English translation, London, 1896.

as the physiological reflection of an already existing class structure of society. Such ideas are found in Aristotle and Plato, with the added notion that to the three sorts of soul (the vegetative, the sensitive, and the rational) there must correspond three groups of men in social life. Shakespeare made use of this doctrine in the opening act of *Coriolanus*, where the First Citizen is compared with society's big toe. Here is an urbane statement of the same idea by a seventeenth-century divine, George Hickes, Dean of Worcester. "Civil equality is morally impossible, because no commonwealth, little or great, can subsist without poor. They are necessary for the establishment of superiority and subjection within human societies, where there must be members of dishonor as well as honor, and some to serve and obey, as well as others to command. The poor are the hands and feet of the body politick . . . who hew the wood and draw the water of the rich. They plow our lands, and dig our quarries and cleanse our streets, nay those who fight our battels in the defence of their country are the poor soldiers. . . . But were all equally rich there would be no subordination, none to command nor none to serve."

There is nothing wrong with the idea that society is an organism. But it is an organism of far higher grade than any of the biological individual organisms, and hence the biologist must even to-day be on his guard against the crude taking over of biological ideas in the service of the ruling class. The Victorians were interested in the analogy. Herbert Spencer, in his *Sociology*, elaborated it in much detail, not even hesitating to refer to the peasant class of a human community as its endoderm-cells, and to the military class as its ectoderm. Executive scribes corresponded to the central nervous system and the king's council to the spinal medulla. All such analogies tremble on the verge of absurdity, and fall to the ground because man in his societies constitutes a higher level of organisation and complexity than anything else met with in the living or non-living world. The possession of consciousness and communication between individuals in a highly developed state, and especially the use of tools in the processes of production, mark off human society as a higher level than biology, just as biology itself deals with systems more highly organised than those of physics and chemistry. Herbert Spencer, however, was aware of the dangers of biological analogies applied to human affairs. So also was Walter Bagehot, in his *Physics and Politics*,[1] which treads on the same ground. And in our own time the physio-

[1] (London, 1872.)

logist W. B. Cannon[1] has shown how the analogy can be safely and fruitfully handled.

Not so certain modern writers, however, such as Morley Roberts, whose book *Bio-Politics*[2] is an outstanding example of the fallacy or distortion here discussed. It will long remain a museum of absurdities. Compare "the action of a national press as a powerful secreting organ needs no demonstration" with "in the organisation of the reticulo-endothelial system it is impossible not to see deep social analogues in the police, the seamen and the soldiery of a nation." Or, still better; "All cells with a nucleus possess tools and weapons with which they do their work. These biochemical tools I have no hesitation in describing as their property. I commend these notions to the legal profession if they should be hard pressed to defend the descent and inheritance of property." (Very candid.) Or again: "Capital is a natural phenomenon and cannot be abolished till we abolish physiology and physics. What begins in nature in the egg will continue so long as eggs are laid." (Loud and prolonged applause from the Right. And all this a hundred years after the work of Marx, Veblen, George, Engels and many others on the analysis of property and capital.) Fascist philosophers such as Spann also argue in this way. But the attempt to justify class robbery, oppression, social injustices and war by conferring upon them unimpeachable scientific authority will never succeed.

A second important case of the distortion of scientific theory for specific political ends is, of course, that of Darwinism. The principle of natural selection necessarily implied a struggle for existence among animal species competing for food and reproductive facilities, and also among the individuals in any one species. It is no criticism of this theory to point out, as Engels did, that it was the reflection on to the animal world of the competitive conditions prevailing in the economic world of nineteenth-century capitalism. In the animal world it happened to be to a large extent true, and the principle of natural selection is to-day held to account very substantially, if not entirely, for the phenomena of organic evolution. The obvious conclusion was that if competitive capitalism was so like the sub-human world, that was

---

[1] *The Body as a Guide to Politics* (London, 1942). He compares the constancy of the physiological internal environment with that social stability in employment and commodity-exchange, combined with individual freedom, which we have not yet attained. See also Science, 1941, **93**, 1.

[2] London, 1938.

just too bad for competitive capitalism. Like the dinosaurs, it was getting a little out of date.

All this, however, was quite beyond the grasp of most Victorian writers and thinkers. All they could see was that the struggle for existence was, as they would have put it, the universal law of life, and that the more red in tooth and claw life could be made in the industrial areas, the more would human civilisation benefit. In other words, the struggle for existence supplied the grandest argument yet available for the necessity of *laissez-faire* capitalism. While it lasted, the going was good, but inconvenient critics such as Kropotkin, in his *Mutual Aid*, demonstrated the large part played in evolution by animal associations including the colonial hymenoptera. Drummond traced the origins of altruism back to the beginnings of the family and the factors involved in primitive reproduction. Darwin himself, though emphasising the competition between species and individuals, had not overlooked the co-operative element. In his *Descent of Man*[1] he said, "Those communities which included the greatest number of the most sympathetic members would flourish best and rear the greatest number of offspring." And as for human society, Marx and Engels, by their historical analysis, showed that capitalism had not always existed, and that there was no reason whatever to think that free competition in exploitation of commodities and labour was more than a stage in man's evolution towards a planned and rationally controlled society.

A third case of systematic distortion and falsification of scientific ideas would be found in the whole history of racialistic theories from Gobineau onwards. While completely contrary to all the most solidly established scientific knowledge about human beings, these theories have played an enormous part in the political life of the twentieth century. We will not further consider them here.

Distortions of science may also occur in the interests of sections of the ruling class. Thus the patent medicine trade and pseudo-scientific advertising come to mind. Professor A. J. Clark's exposure[2] of the former in 1938 revealed the colossal annual financial turnover which the business of battening on the people's ill-health achieves; at that time about equivalent to the national annual public expenditure on hospitals. Exposures, however, are few, since the patent medicine trade constitutes the greatest single advertiser in the daily newspapers, and is therefore in a position to exert overriding influence against the

---

[1] Ch. 4, p. 163.   [2] "Patent Medicines," Fact, 1938, No. 14.

publication of any criticism, no matter on what scientific authority. I say this on the basis of first-hand editorial experience. Furthermore, it is not generally realised what systematic mystification goes on within important departments of national life such as the chemical industry. A friend of mine who became an industrial chemist used to tell me of the code employed by the storekeepers in his firm, in which every chemical substance which the workmen had to draw from the store was given a wrong, and if possible, a confusing, name, so as to avoid at all costs the danger of any trade secret leaking out to a competing firm through the workers employed. On one occasion some plumbers who were doing a job in the works and drew what they were told was "rosin" from the store, drew something very different, with remarkable effects. All these phenomena are not those of a healthy society.

It is clear, therefore, that one important sense of the phrase "keeping science pure" is that scientific men should be constantly on their guard against the distortion of truth for reactionary, political or economic purposes. I write "reactionary" deliberately. Progressive political movements have no need to distort it.

The Position of the Scientist in Society.

We return at last, then, to the position of the scientist in society. He is not only a scientist, but a citizen as well. There has been much talk (and not uncommonly, in the editorials of Nature), about the importance of giving the scientist more say in government. The value of this is not to be denied. But when it is carried further, as in the views of the "technocrats," and when it is suggested that scientists should themselves be the rulers of society, the suggestion reveals its superficiality.[1] What has to be decided is the ultimate end for which society exists. Is it the greatest happiness of the greatest number? The right of every man and woman to life, liberty, and the pursuit of happiness? The right of the individual to work, food, love, and opportunity to develop his talents in the common service? Or the military domination of one state or "race" over other peoples, the

[1] This was the theme of another controversy in Nature. R. H. F. Finlay (1941, **147,** 119) expressed the hope that "scientific instead of party government would prove the unifier of man, and usher in an era of universal peace." The distinguished entomologist, V. B. Wigglesworth (1941, **147,** 206) hastened to point out that divided counsels are the very breath of science, which is always growing. It would be a tyranny indeed to give to the current beliefs of science the force of law.

perpetuation of a class supremacy and privilege, the regimentation of a mass of helots for the upkeep of a veneer of oligarchic culture? The scientist, as administrator, can make administration more efficient. He cannot take away from humanity the duty of facing up to the fundamental moral principles. He cannot relieve humanity of the fundamental moral choice. Only by political action can moral choices be implemented and political history made.

Science must be set apart from all other beliefs as possessing a peculiar holiness, wrote Polanyi in the passage quoted above. The word "holiness" here is really significant. To the court of religion he has appealed, to the court of religion he shall go.

In one of the most brilliant discussions of the history of religious ideas known to me, John Lewis described in some detail how throughout the history of jewish and christian theology there has been unceasing strife between the sacred and the secular.[1] Prophets arise to castigate the shortcomings of the social life of the people, to menace the wealthy exploiters with divine wrath; to demand restitution of property robbed from the workers, the fatherless and the widows; and to turn the financiers out of the Temple with a whip of short cords. But repentance is soon over, the existing economic processes resume their sway, and the religious ideas of love and equity, never wholly lost, are stored up in churches under the guardianship of priests. "Religious emotion," wrote Marx,[2] "is, on the one hand, the expression of actual misery, and on the other, a protest against actual misery. Religion is the sigh of the oppressed creature, the kindliness of a heartless world, the spirit of unspiritual conditions. It is the people's opium. The removal of religion as the illusory happiness of the people is the demand for its real happiness. The demand that it should give up illusions about its real conditions is the demand that it should give up the conditions which make illusions necessary. Criticism of religion is therefore at heart a criticism of the vale of misery for which religion is the promised vision."

Criticism of the vale of misery, carried through to its ultimate point, becomes the demand for a thorough renewal and reconstruction of society in the interests of rationality and love, interests which cannot rule under the system of exploitation of man by man in social classes. When once ownership of the means of production and of all

---

[1] "Communism the Heir to the Christian Tradition" in *Christianity and the Social Revolution* (London, 1935).
[2] *Introduction to a Critique of Hegel's Philosophy of Law*.

natural resources is wrested from a class and transferred to the community, "the relations between human beings in practical everyday life would assume the aspect of perfectly intelligible and reasonable relations as between man and man, and as between man and nature."

"The moment this becomes possible," John Lewis goes on, "the religious institution is confronted with an altogether new crisis. Hitherto it has stood for an *unrealisable ideal*, and has alternated between secularisation and transcendentalism. It now becomes possible to enmesh the ideal in the material world without loss.[1] So long as the social and economic organisation was of such a character that it could not permit the realisation of his ideals, the idealist was steadily forced to accommodate his principles to its inexorable demands in so far as he decided to live and work in society and not to dream. On the other hand, if he determined to keep his ideals intact, then he could not effectively grapple with reality, and was compelled to become a hypocrite (making the best of both worlds) or a mystic. But when social development reaches the stage when ideals are realisable, the struggle becomes capable of a successful issue. It is not settled, but it is no longer condemned to futility by the very nature of the conditions. As a consequence, the whole structure of religion changes. The ecclesiastical, devotional, and mystical forms proper to a dualistic period, become obsolete. Religion has been adapted to the needs of a class society; it must now suffer complete transformation as the classless world approaches. To some this spells the death of religion, and 'blank materialism;' to others it is what they had always sought. The prophet should be able to welcome the new age. Now at last it will be possible to manifest the will of God in social relations, and show forth the glory of His purpose in the common ways of life. In the words of Zechariah[2] and Ezekiel:[3]

> 'In that day there shall be written upon the bells of the horses, Holy unto the Lord. . . . Yea, every pot in Jerusalem shall be holy unto the Lord of Hosts.'

[1] The belief that it is *possible* to enmesh the ideal within the real world, to transform the real world, is profoundly Christian, but also profoundly Hebrew and profoundly Confucian. In ancient Chinese writings the sages are frequently said to unite Heaven and Earth by their virtue (teh, 德). Apart from magic undertones, this surely means that by their insight into the nature of human community, they understood how to enmesh the highest moral good in the real world. As the *Ta-shioh* says, one must find the highest moral good and then stop (chih, 止) there, not going on beyond it or evading it by sophistical arguments.

[2] xiv. 20, 21.   [3] xliii. 12.

'This is the law of the house . . . the whole limit thereof round about shall be most holy. Behold, this is the law of the house.'

"The sacred is far from being the 'wholly other'; it is the quality of the secular raised to its highest power and consecrated to the noblest purposes. As each part of life is integrated into the social organism, it finds itself, and takes on the special quality that belongs to a part of a new whole. On the other hand, transcendentalism withers away. It is no longer necessary to project into another world the order, the justice, and the beauty, which we cannot achieve in this. The life process of society loses its veil of mystery when it becomes a process carried on by a free association of producers, under their conscious and purposive control."

I make no apology for quoting this splendid passage at length, since it perfectly gives the judgment for which we were looking. It embodies the profoundly christian doctrine that the world *is* redeemable. There is nothing holy but flesh and blood. The essence of our purpose should be, not so much, as M. Polanyi would have it, to keep science "holy," but to make the whole of human society holy. This is only another way of pointing to the high levels of social organisation to which humanity has yet to climb. At those high levels scientific truth will be well able to look after itself, since on the foundation of scientific truth alone can those high levels be built. The scientist, it was said, is not only a scientist, he is a citizen as well. He must be a citizen, not only of that temporal nation to which he happens to belong to-day, but also a citizen of no mean city, the *Civitas Dei*, or, if you prefer to call it so, the World Co-operative Commonwealth.

Thoughts of a Young Scientist on the Testament of an Elder One
(John Scott Haldane)

(Based upon an address to the World Congress of Faiths, 1936; later expanded for Fact; and Science and Society, 1937)

ALL through this century and for some time before it began, the meetings of the Physiological Society were enlivened and adorned by the presence of a peculiar stooping figure, whose bushy eyebrows and moustache hardly concealed a countenance of great charm and originality. It was John Scott Haldane, one of England's best physiologists, a tireless investigator of the mechanism of respiration, an acknowledged expert on the conditions of labour in coal-mines, and a philosopher whose permanent place in the history of biological theory was long ago assured. Personal contact with him I had none, but from my earliest undergraduate days I had been nourished on his work, and warned against his errors by those who did not agree with him. Those of us who later made a special study of biological theory had to pass through his books, and, as it were, out at the other side, before reaching a satisfactory viewpoint. In this way, his *Mechanism, Life and Personality*, his *Materialism* and his *Philosophical Basis of Biology* were all important books, and like those of Hans Driesch, had to be fully digested before we earned the right to abjure for ourselves the name of vitalist.

In 1936, in his 77th year, he died. But some time before, he had been invited to take part in a World Congress of Faiths by contributing a paper on science and religion, and this paper he wrote. It was to be his last writing. And it fell to me to open the discussion on this paper at the Congress itself, an undertaking which could not be refused, partly in gratitude for the violent intellectual shocks which I had received for so many years before from J. S. Haldane, partly on account of the stimulating nature of this "testament" itself. If, as will be seen by those who read further, I disagreed profoundly with what he said, this disagreement was to some extent a disagreement of generations, for the generation of to-day cannot accept the formulations of the past. Too much has happened since the quiet days of

Haldane's pre-1914 Oxford. And it seemed to me that to discuss materialism now in much the same way as it would have been discussed thirty or even sixty years ago was just unrealistic.

Religion and the Forms of Materialism.

Even at the first paragraph I was brought up sharply.[1] "To many persons in modern times," he began, "it seems as if the only reality is what can be interpreted in terms of the physical sciences, with the addition, however, that certain physical processes occurring in the brain are mysteriously accompanied by consciousness, the quality of which depends on the nature of these processes. This belief is known as materialism, and for those holding it, religion is necessarily no more than an illusion based on ignorance." "This belief is known as materialism." But to speak in this way is to make no distinction between materialisms, of which there may be more than one. Most of Haldane's criticism bore against traditional mechanistic or metaphysical materialism only. Dialectical materialism he in no way considered.

The character of mechanical materialism is, I suppose, the belief that all changes in living and non-living objects are ultimately reducible to changes in the position of invisible particles which simply *are*.[2] These particles and their motions are quite independent of our thought. The degrees of complicatedness in the world are illusory, or if not illusory, are degrees of complexity only, not of organisation. All should be reducible to atomic laws. But the essence of dialectical materialism, on the contrary, is the acceptance of the existence of diverse levels of complexity and organisation, and the interpretation of them as successive stages of a world-process the nature of which is synthetic or dialectical. Order and organisation are fully allowed for.

As far as metaphysics is concerned, many scientists have always felt a strong disinclination to take up a position in the classical philosophical controversy between realism which asserts the primacy of the object, and idealism which asserts the primacy of mental cognition. This controversy seemed uninteresting and academic. They were loath to make the old sharp distinction of the philosophers between the world of spirit and the world of matter. After all, for the biologist there is no strict separation. In animal and human behaviour there is a

[1] The full text of J. S. Haldane's paper will be found in the *Proceedings of the World Congress of Faiths*, 1936.
[2] L. J. Russell's phrase, Aristotelian Society Symposium, 1928.

unity; the "mental event" cannot really be separated from the "physical neural event," and it is profitless to try to do so.[1] So in the coming into being of the world as a whole, we should envisage a unity, as dialectical materialism in fact does. Out of original chaos a vast flowering of the new has originated, and that is all that can be said. This point of view can be found already in scholasticism, when Duns Scotus asked "whether it was impossible for matter to think?"[2] In the seventeenth century Hobbes maintained that it was impossible to separate matter from matter which thinks. In the eighteenth, we have that celebrated dialogue of Diderot with d'Alembert.[3] They discuss the development of the sensibility of the chick embryo in its egg.[4] Diderot maintains that one must either admit some "hidden element" in the egg, penetrating into it in some unknown way at a certain stage of its development, an element about which we know nothing, whether it occupies space, whether it is material, or whether it is created independently for each chick (an idea contrary to common sense and leading to inconsistencies and absurdities)—or one must make a "simple supposition which explains everything, namely, that the faculty of sensation is a general property of matter, a product of [certain forms of] its organisation." To d'Alembert's objection that such a supposition implies a quality which is essentially incompatible with matter, Diderot retorts: "And how do you know that the faculty of sensation is essentially incompatible with matter? You do not know the essence of anything, neither of matter nor of sensation."

Dialectical materialism itself may, in a sense, be considered historically as a dialectical synthesis. Classical metaphysical materialism and idealism were the antitheses which gave rise to a deadlock. The former was unable to account for any of the higher manifestations of the human spirit, values, artistic creation, altruism, love. It had

[1] Cf. Pavlov's famous essay on the fusion of the subjective and the objective. Introspective psychology plus the physiology of conditioned reflexes and other neurological methods promise at last an understanding of the relations between the mental and physical levels of organisation (*Lectures on Conditioned Reflexes*, London, Vol. I, p. 39, Vol. II, p. 71). Pavlov had been stimulated by the brilliant insight of a Russian physiologist of an older generation, I. M. Sechenov, whose *Cerebral Reflexes* appeared in 1863 (see Sechenov's Selected Works, English edition, Moscow, 1935).

[2] Cf. Engels' exposition of this in *Ludwig Feuerbach*, p. 84.

[3] Cf. Lenin's exposition of this in *Materialism and Empirio-Criticism*, p. 104.

[4] The "problem" of the entry of the soul into the embryo is of great antiquity; see the account of it in my *History of Embryology*. For an insight into the way in which the development of behaviour in the embryo is treated by modern science, the papers of Kuo Zing-Yang in the psychological and physiological literature should be consulted.

to be "ascetic" or "misanthropic" in picturing its grey world of clashing particles. Idealism, on the other hand, always verging on subjectivism and solipsism, could never be united with the requirements of the scientific mind or with progressive social movements. The difficulty had occurred long before to the scientific philosophers of the ancient Mediterranean world, to Epicurus and Lucretius, who introduced the famous "swerve" or *clinamen* of their atoms in order to save the possibility of free will and other complex phenomena of human life and society. This device was rather a transparent one, and indeed no true synthetic way out of the difficulty was available until the fact of cosmic, biological and sociological evolution came to be appreciated by mankind. Room could then be found for the understanding of all the highest phenomena of human life as the characteristic properties of the highest levels of organisation in the known universe, without at the same time having to give up materialist philosophy, i.e. the conviction that, though there may be a certain distortion in the process, external nature does exist independently of our observation and that we do receive a substantially true picture of it as we investigate it. It did, moreover, exist for many aeons without the contemplation of any observing spirits such as our own. Their appearance was part of the formation of ever higher levels of organisation and complexity.

These new levels were to some extent recognised by Haldane himself. Thus later on he wrote, "Vitalist biologists assumed unjustifiably that in a living organism something interferes from outside with physical processes. For the newer biology there is no interference from outside, but the integration characteristic of life is inherent in the events perceived, and they cannot be described apart from it." Indeed, Haldane's great service to biological theory was the way in which he persistently called attention to the special form of organisation existing in living things. On the other hand, his great failure consisted in his defeatist wish to accept this principle of organisation as axiomatic, instead of tracing its relation to the lower principles of organisation seen in para-crystals, colloids, and similar states of matter.

Again, for materialists, Haldane wrote, "religion is necessarily no more than an illusion based on ignorance." For me this was frankly incomprehensible. His conception of religion must have been very rationalistic, some kind of theistic *explanation* of the universe. For me, on the contrary, ever since I first read Rudolf Otto's fundamental book *The Idea of the Holy*,[1] it has always seemed that what distinguished religion from theology or inferior philosophy was the sense of the numinous,[2] the divination of sacredness in certain external things, persons or actions. At first this sense is fetishistic. It attaches purely to certain objects, apparently at haphazard. Then great pioneers discover that certain actions are holy—comforting the fatherless and the widow, or putting down the mighty from their seat and raising up them that fall, or following justice and mercy, or above all, the ἀγαπή τοῦ πλησίον of the Gospels, the love of one's comrade.

This kind of religion seems to me to have no metaphysical commitments of any sort. Whether your comrade is a higher ape with a very complicated nervous system embodying a vast set of reflexes, or on the other hand, a non-material soul inhabiting in some curious way a material body—he is at any rate both beautiful and lovable, and in a certain sense pathetic, and the disposition to love him is a grace not affected by your theories of his nature.[3] I do not know what is meant by the phrase the pastoral theologians use, "the love of souls," but I have found that superficial animosities and differences tend to disappear if men and women are visualised as the children they once visibly were. Some such deeper comprehension and love must have informed the carver of one of the greatest of all statues of Mary, Conrad of Einbeck, whose work is still in the great church at Halle.

---

[1] English translation by J. W. Harvey (Oxford 1923).

[2] From the Latin word *numen*, which means a deity, Otto coined the adjective *numinous* to designate the quality of sacredness or holiness attached to things, persons or actions. In mediaeval christianity, the paten and chalice in the Mass were thought to possess this quality, so that even today the subdeacon, being a layman, may only handle them ceremonially wearing the "humeral veil." This is akin to the "mana" of the anthropologists. But persons and their actions may also be numinous, for example, the widow in the Gospels contributing her small coin to the collection, or Irenaeus saying on the morning of his martyrdom, "Now I begin to live," or the death of the five Leveller Corporals in Burford Churchyard, or Dimitrov's immortal witness to certain principles in the face of dementia enthroned.

[3] After I had written the above words, I came across the following in S. Alexander's *Space, Time and Deity* (London, 1927), Vol. II, p. 32:

"The experience which assures us not inferentially but directly of other minds is a very simple and familiar one, that of sociality, and it has a double aspect. Our fellow human beings excite in us the social or gregarious instinct, and to feel socially towards another being is to be assured that it is something like ourselves. We do not first apprehend that another being is a mind and then respond to him, whether positively in affection or negatively in aversion; but in our tenderness or dislike we are aware of him as like ourselves... But we do not experience the satisfaction of sociality till the creature towards which we act socially reciprocates our action."

Cf. also Henry Maudsley's *Body and Will* (London, 1883), p. 45.

The Development of Religious Feeling.

More emphasis ought, indeed, in justice to traditional christianity, to be laid on the materialist elements in it;[1] for example, the sacramental principle. In the poetic symbolism of the sacrifice of the Eucharist, we take clearly materialistic things, bread and wine, and with them we offer and make, to a God anciently conceived of as both immanent and transcendent, a holy and efficient sacrifice, remembering that (in this language) material objects are necessary as carriers of grace. Grace can have no existence in isolation. If we love not our brother whom we have seen, how can we love God whom we have not seen? The material bread and wine of this sacrifice ought to teach us that only by caring for the body and blood and secular spirit of our comrade can we assist his soul, whatever that may be. The material bread and wine ought to remind us that the most exalted spiritual things are connected with, and have arisen in evolution out of, the most primitive processes of living and dead matter. Christians themselves rarely seem to understand this.[2]

No metaphysical commitments. Haldane's fundamental error of identifying religion with metaphysical idealism made him blind to one of the most impressive facts of our time, namely that everywhere there is proceeding a persecution paralleled only by those of the early Church, a persecution of materialists who hold this doctrine of love of our comrade in a more thoroughgoing and enlightened manner than it has even been held before. Of course, I refer to the communists, and those who sympathise with them, who in all countries, Latin, Slavonic, Anglo-Saxon, are being imprisoned, tortured, and murdered every day.[3] Their doctrines might be described, perhaps rather provocatively, as the highest form which religion has yet taken, a form in which it negates itself, and must necessarily be at war with all previous forms of itself. In my view, the communist has, although in most cases he will never admit it, a more highly developed sense of the holy than any of the adherents of traditional religions, for he sees that the oppression of man by man is unholy, and he is determined to banish it from the

[1] Cf. McTaggart, J., *Some Dogmas of Religion* (London, 1906), p. 250.

[2] But long after I had written this, I found that Archbishop Wm. Temple does. In his Gifford lectures (*Nature, Man and God*, London, 1934), he says "Christianity is the most avowedly materialist of all the great religions" (p. 478). And later he approximates his own position to that of dialectical materialism (pp. 487 ff.).

[3] Cf. A. D. Nock, *Conversion* (Oxford, 1933), p. 228, where the position of early christians is likened to that of modern communists.

world.[1] In the effort to do this, he is perfectly realistic. For example, he understands the necessity for making the contacts between men and machinery beneficent, not lethal. This is the kind of ethic appropriate to our civilisation. He has faith also, the faith that infinite capacities for good are resident in man, the faith that the early christians had, that the *Regnum Dei* can be built on earth. Alone in the world to-day he has noted the apostle's warning: "He that despiseth man, despiseth not man, but God." Finally, his philosophy is precisely dialectical materialism, the view that one original creative event, probably for ever impenetrable to us and therefore hardly worth prolonged discussion, gave rise to a succession in time of dialectical developments, ever higher stages of organisation being reached. The classless state of justice and happiness on earth itself forms part of this succession and belief in it is therefore no mere desperate act of faith, but a part of an eminently rational philosophy and a declaration of unshakeable confidence—"Magna est Veritas, et praevalebit."

The highest form which religion has yet taken is a form in which it negates itself, and must war with all previous forms of itself. These words require further elucidation. For this purpose I suggest that instead of defining religion, with Haldane, as a kind of theistic philosophy without historical antecedents, we should accept the following propositions. First, that as Otto showed, religion begins in shuddering fear and dread of the numinous, a special category of external nature; and is then later attached to ethical actions with exceedingly powerful psychological means by certain persons, such as Jesus, whose death on the cross epitomises all the guiltless sufferings of the just. Secondly, as would follow from the conceptions of Marx and Engels, this numinous quality was from the beginning also attached to the exploiting lord as opposed to the exploited serf. Hence all organised religion tended to stabilise the exploitation of man by man. In so general a discussion as this, examples will occur to every reader, but mention may be made of the sacredness of the Roman imperium,[2] the history of the Papacy, and St. Paul's phrase "the powers that be are ordained of God," to say nothing of such minor phenomena as the deacon's dalmatic placed upon the Kings

[1] The numinous shifts as the economic possibilities open out during human social evolution. Those who are ahead of their time are those who visualise the new possibilities and are able to divine the new position of the numinous.

[2] Cf. the extraordinary forms which this *sacratissimum ministerium* took in the time of the thirteenth century Hohenstaufen Emperors (*Frederick the Second*, by E. Kantorowicz, London, 1931).

of England at their coronations. Religion, like Folk Art, is the spiritual protest of the oppressed creature, and organised religion is an opiate in so far as it canalises and confines this protest purely to the spiritual realm, turning it thus into an escape from the real.

Thus religion begins with fear; is stabilised by priests as an instrument of subjection; is transformed by prophets into ever higher forms of the sense of the holy, and ends in the idea of love as holy. By removing earthly oppression, the necessity for spiritual compensations is also largely removed, and in this way when perfect love casts out fear, religion (in one sense) goes with it. In another sense it remains as the symbolism of comradeship. Hence the paradox of communism—the only persons in the world to-day who take the Gospels seriously are just those who declare themselves the enemies of all religion. This fact, I cannot help feeling, might be more congenial to the central figure of the Gospels than most christians would imagine. For when that which is perfect is come (the Kingdom of God), that which is imperfect (organised religion) shall be done away. Or, as John Lewis has put it,[1] Religion must die to be born again as the holy spirit of a righteous social order.

The Parallel of Folk Art.

Further mention of folk art at this point is no digression. As the semi-inarticulate manifestation of the consciousness of an oppressed class, it demands comparison with religion. And here also the sacramental principle is evident; the actors in the mummers' play are saying far more than they mean, the sword-dancers and the wren-hunters are symbolic to the core. Nor is this remote from Haldane, writing in his study his last contribution, for no one knew better the coalfields than he, and no one could have watched the rapper sword-dancers of Northumberland and Durham with greater sympathy. But the truth has never yet been told about folk-verse, folk-music, folk-dance. The folk-songs of England are the songs of the peasantry (labourers, cowmen, woodmen, shepherds, etc.), the seamen (in the form of chantys), and other types such as the town artisan and tradesman, the peddlar, and the rank-and-file of the army. And two features of working-class life especially have stamped their character upon the folk-songs, first the lack of education of the people, and second the hardness of their life and the lack of security in their occupations. The second is the more important. In the haunting minor tunes, the

[1] In *Christianity and the Social Revolution* (London, 1935).

strong preoccupation with the evanescence and sadness of life, and the fundamental seriousness which makes folk-songs so beautiful as well as so charming, we see the expression of the people's sufferings. Even the apparently most individual theme of love is in no way immune from this, for whereas in its simplest forms it may involve great torment, all this pain is enormously increased by adverse economic and social conditions. Hence the ballads of violent action, the songs of separation, and the fewness of the songs with happy endings. It is in this light that we can recognise the essential rottenness of a pseudo-folk book such as *Mr. Weston's Good Wine*, for although to the folk, the standing conventions of capitalist society, such as the persecution of unmarried mothers, appeared as unalterable laws of the universe; a sophisticated author well knows that they are nothing of the kind, and yet in spite of that, fails to indicate their true nature. There are many folk-songs about the kidnapping of lovers by the pressgangs, but it would have been fantastic to accept this institution as inevitable.

Another point to which attention is not usually drawn is the affection of the folk for outlaws such as Robin Hood or the innumerable highwaymen who appear in the songs. Whether real or legendary, these figures are important indications of the psychology of expropriation on the part of the country working-man, and they exist everywhere. Not only in England, but also in the Carpathians, does Robin Hood, under the name of Janosik there, waylay the rich knights to the joy of the Polish and Slovak mountaineers. Listen to the verse from "The Robber":

> "I never robbed a poor man yet,
>   Nor never was I in a tradesman's debt,
> But I robbed the lords and the ladies gay
>   And took the gold,
> And took the gold to my love straightway."

The same psychology also appears in the way the folk took to heart the biblical story of Dives and Lazarus, as shown in the famous song "Lazarus and Diverus."

Folk-song, like religion, is the sigh of the oppressed creature. The disappearance of folk-song in the past was partly due to active suppression. Alfred Williams, in his *Folk-Songs of the Upper Thames*,[1]

[1] (London, 1923.)

says that the police, who were apparently unable to distinguish between a folk-song and a "rough house," exerted themselves about 1900 to forbid all singing in inns and public-houses. In the future, its disappearance may be one of the prices we shall have to pay for a more general happiness and well-being. But we may reflect that when the oppression of man by man has been liquidated, a certain number of fundamental limitations will still remain. Inhabitants of space and time, subject still to the oppression of mortality, men will surely still be able to create and appreciate the kind of art-productions—for such they are, however communal and however artless—which we know as folk-songs.[1]

The folk-dance, too, has interest both for the archaeologist, since it is found in association with many different rites, often undoubtedly the late survivals of pre-christian festivals; and also for the amateur of dancing, since in many cases it is skilled and beautiful in the highest degree.[2] But even here we cannot fail to note the same sociological factor. The seasonal festival was a recognised means of escape from the state of subjection in which the peasant normally lived. "We daren't come out but on Plough Monday," an East Anglian dancer said to me once, "they'd have the law on us if we ploughed up a doorstep, but on that day they can't touch us, because it's an old charter." Unfortunately, even this was often denied. In 1816 the ringleader of a gang of "plough-bullocks" was severely reprimanded and fined before a magistrate before whom he was brought. "These men had a notion that they had a privileged right on certain days in spring, to exact donations from respectable residents, and in default of payment to damage their premises." The authentic bourgeois tone. The two great theophanies of the bourgeois, the Puritan and the Business Man, both hated the Morris, the Maypole, the Plough-Stots, and the Mummers. They wanted to make the world safe for the profit of godly industry, and they succeeded, although with time, it became, as we have observed, much less godly and distinctly more profitable.

In the socialist state of the future, as in the Soviet Union to-day, the traditional ritual dances of the people will be treasured indeed. They are the pure creations of the working-class, and they will

[1] The justice of this belief is strongly reinforced by the interesting article of M. Azadovsky on the present position of folk-songs in the Soviet Union (*Lenin Commemoration Volume*, Academy of Sciences, Moscow, 1934).
[2] Cf. my monograph "Geographical Distribution of English Ceremonial Folk-Dances" in Journ. Eng. Folk Dance and Song Soc., 1936, **3**, 1.

unite by a remarkable continuity the developed communism of the future with the antique primitive communism of the far past.

Modern Manichaeism.

All this is a manifestation of the consciousness that what ought to be is distinct from what is. In a fine passage, Haldane comments on this. "However what is evil, or what ought not to be, is interpreted, we find that the distinction between what ought, and what ought not, to be, is generally acknowledged, together with the obligation to further what ought to be. It is on the recognition of this obligation that the religions of the world are founded; and although they differ in matter of detail, they are united in their recognition of what is good, or what ought to be, and the obligation to further it. The belief, however, in spiritual powers of evil is now, I think, a waning one in all civilisations; and evil has come, or is rapidly coming, to be regarded commonly as the unchecked prevalence of what is interpreted as belonging to the physical world over what is spiritual." A good beginning but a bad ending. I pardoned myself for the fleeting thought that not all living religions only, but also all the dead ones, were to be represented at the Congress. The Manichaeans themselves could hardly have put their case better.[1] Such a conclusion, violently at variance with christianity and communism alike, could only lead, it genuinely followed through, to a retrograde asceticism. For marxism, the origin of evil, like the ultimate origin of everything else is, I suppose, inscrutable, and the problem of evil is not a problem of fixing the responsibility on some one, some being, or some constituent of the universe, but a problem of biological engineering. "Philosophers have talked about the world enough, the time has come to change it." We must, of course, accept in all probability an irreducible minimum of pain and sorrow so long as man is a patient being immersed in space-time, but this is like that irreducible minimum of the alogical,[2] to which scientific explanation is always asymptotically tending. The reduction of the alogical in our picture of the universe is what Haldane

---

[1] Manichaeism was one of the many religions which disputed with christianity for the mastery of the late Roman world. Its founder, Mani, maintained that matter was utterly and irredeemably evil. Its adepts were, therefore, under the necessity of pretending not to eat or to perform any other bodily functions (see *The Religion of the Manichees* by F. C. Burkitt, Cambridge, 1925).

[2] Tennant, F. R., *Philosophical Theology* (Cambridge, 1928), Vol. I, ch. 13. Dingle calls it *Deus ex aequatione* with the character of an arbitrary constant.

called, in the manner of his generation, the search for truth. On it he bases his argument for the existence of God.

"The search after truth," he writes, "which appeals to all man, implies the existence of personality which extends over and includes all individual personalities. In other words, it implies all-embracing personality, which for religious interpretation is the personality of God. In the furtherance of truth, as revealed in experience of any kind, God as the supreme Person is revealed to us, and our trust in experience is trust in God." But this is more poetical than philosophical. If transcendence is assumed it is unconvincing. If immanence only is meant, why should not this all-embracing personality be that of the human collectivity, as McTaggart[1] and Marx alike would require? If, moreover, there is any value in Otto's theory of religion, "religious interpretation" is a phrase almost without meaning. Religion is not concerned with interpretations, but with emotions favourable or antagonistic to social coherence. Emotions favourable to social coherence are perhaps the analogues of those bonds which hold entities together at the physical level. From this point of view, one can only accept the word "God" as a poetic term analogous to that used in the apostolic precept already referred to—"He that despiseth man, despiseth not man but God," i.e. the highest worshipful values we know.

The Ethics of a Machine Age.

To some it may seem that this criticism of Haldane's paper has too political a character, but one must remember that it has been the opinion of many a christian theologian that politics cannot be divorced from ethics; is, indeed a branch of ethics. Since ethics, in its turn, cannot be divorced either from religion or from philosophy, no one can undertake to discuss materialism and religion at the present time without considering dialectical materialism and the communist paradox within the realm of the numinous. In his address to the World Congress of Faiths, Haldane made no mention of any political issue. But in one of the pronouncements of this remarkable man, he did touch upon such problems, and it is interesting to compare what he then said with the theme of his last paper. It was in 1924 that he was invited, because of his long life of research in industrial physiology,

[1] J. McTaggart, *Some Dogmas of Religion* (London, 1906), pp. 12, 121, 133; "If all reality is a harmonious system of selves, it is perhaps sufficiently godlike to dispense with a God," p. 250.

to become President of the Institution of Mining Engineers, and the presidential address which he gave on that occasion was reprinted in his book of essays, *Materialism*, eight years later. After some introductory paragraphs about the capital value of British coal-mines, in which he showed a naïve acceptance of the whole usurious system of shareholders and owners, he proceeded to develop a peculiar argument. Modern materialism, he felt, was at the bottom of the failure of the capitalist system to work properly. Modern materialism involves the tacit assumption that everything, including the economic system, works mechanically, hence the industrial machine of a coal-mine is conceived of as a "soulless" interplay of forces. The corollary is that pitmen are not thought of as human beings but simply as "hands," and are to be asked to work as long hours as possible for as small a wage as they can be induced to accept.

If this is "materialism," there is certainly nothing good to be said for it. But has it anything to do with the philosophical debate? Is it not rather a question of ethics and morality incarnate in problems of practical economics? The question is whether the welfare of the workers, or the accumulation of profits, is the primary consideration. If profit-making is by far the most important consideration, as under capitalism it must be, then all that has to be done is to hand the factory over to the investigations of "efficiency engineers," who will plan for maximum production irrespective of anything else. Care for the workers will take the form of pushing "rationalisation" up to the very maximum limits of what the "hands" will stand. Thus the conveyor belt (satirised in Chaplin's famous film *Modern Times*) has increased both drudgery and tension. Most jobs simple enough to be done in series can be done entirely mechanically, but where labour is cheap and conditions need not be considered, it is naturally not worth while making the machines. Few people yet realise the marvels of self-acting productive machinery which the linking of "receptor engines" such as the selenium cell and the infra-red ray receiver,[1] with the "effector engines," could achieve. This was the moral of another famous film (whether understood or not by the audiences which saw it)—René Clair's *A Nous la Liberté*, which ends with the gramophone factory turning out the gramophones almost entirely automatically, while the factory staff sit on the banks of the adjacent canal enjoying the fishing. Under this symbolism a profound truth was

[1] See R. Calder's article "Millions of Men with Teaspoons" in New Statesman, 1940, p. 178.

concealed. Not only abundance of all commodities could be attained by the proper use of the world's raw materials and the available technical and mechanical genius, but also abundance of leisure. But the ruling classes of the world are afraid of leisure for the people. Just as up to the 18th century free thought was regarded as the preserve of the ruling classes, so now leisure is a jealously guarded privilege. Under conditions of greater leisure, the workers might begin to think.

But to return to the ethics of modern technology. As Stuart Chase pointed out in his remarkable book *Men and Machines*, the contacts of men with machinery may be classified into a list. Some contacts are beneficial, others are lethal. A list of this kind (modified from Chase's presentation), would be as follows:—

Beneficial Contacts:
(1) Operating machines with a large measure of individual responsibility (e.g. locomotives, motor-cars).
(2) Operating stationary machines with responsibility only for speed or direction control (e.g. liner engines, turret-lathes).
(3) Inventing, designing, repairing, or inspecting machines.
(4) Playing with machines (e.g. amateur radio, models).
(5) Submitting to a machine process in someone else's control (e.g. dentist).

Neutral Contact:
(1) Being carried by a machine with no responsibility for control (e.g. train, ship, plane).

Lethal Contacts:
(1) Tending machines with no responsibility for control (e.g. feeding raw material, receiving and stacking finished products).
(2) Participating in the work of machines with no responsibility for control (e.g. performing comparatively simple operations on objects passing on a conveyor belt).
(3) Submitting to a machine process in someone else's control (e.g. being shelled or bombed).
(4) Forming part of a machine under the control of others (e.g. army or prison "discipline").

These are the kind of considerations which should be taken into account in thinking of productive labour. Under the capitalist economic system and the old liberal conception of the freedom of the worker to sell his labour in the market for whatever wages, such contacts are not classified. Whether they are beneficial or lethal does not matter in capitalist society. Haldane might well have considered this.

But it was the direction of his thought. His good instinct was shown as follows:—"It is not with scientific abstractions called 'labour' and 'capital' that British mining engineers have to work, but with their fellow-countrymen, their own flesh and blood. These fellow-countrymen will give loyal and efficient service, will face any danger, will forgive real or imagined mistakes, and will take the rough with the smooth, the bad times with the good; but what they will not tolerate is being treated as if they were mere tools, to be cast aside without compunction." Haldane went on to speak of the comradeship which often existed between all ranks in the First World War, and of the "comradeship taught in the Gospels." His good instinct was clearly inadequately backed by political insight. All the remedy he could propose for the evil effects of his materialism was "effective and sympathetic contact between the head management and the men employed." No glimmering of an idea that there might be something wrong about the basic relationship of master and man, of owner and labourer. No suggestion that such relationships belong to the category of exploitation, that productive machinery should not be in private hands at all. No appreciation of the vast difference in material and spiritual goods, advantages and privileges, which separates the mine-owner from the pitman. So deeply engrained in a middle-class scholar are the convictions of upbringing, the belief that classes are a natural phenomenon as inevitable as the tides or the weather. And we see how the natural instincts of so admirable a man as Haldane lead to a philosophy quite suitable for fascism. His recommendation that direct contact between master and man was the main need of our time is a piece of mediaevalism, a retrograde step, an invitation to fascism because it plainly conserves the stratification of classes. Back to the master-craftsman and his journeymen subordinates. The air resounds with such slogans. Back, said Sir Josiah Stamp (at a British Association meeting) to a lesser production of goods and inventions.[1] A halt must be called to science. Back to the pre-war period, said the German Nationalists. Back beyond the Renaissance, say Spann and the "German

[1] Qualified a good deal, it is true, in his later *Science of Social Adjustment* (London, 1937).

Christians." Back to primitive paganism, says Klages. The essay of Karl Polanyi in *Christianity and the Social Revolution* is a perfect museum of such regressive movements. One may well feel depressed.

Then one's eye catches the following, written without doubt in all sincerity—"If at any time we feel discouraged, we should remember that behind our comradeship there stands the wider comradeship of which the British flag is for us the symbol—a symbol alike of righteous and humane dealing, and of the fearless power that lies behind it. Let us never allow the fickle waves of passing public sentiment to obscure our vision of that wider comradeship." How the women and children immolated in the early industrial age in the pits of those mining districts that Haldane knew so well would have reacted to such words, it is hard to say. The Irish victims of the Black-and-Tan regime, the cheated native miners of South Africa, the Chinese mandarins vainly trying to stem the import of opium into China by British trade and finally forced to fight unequal wars over it, the Indians struggling against a rule of censorship and concentration-camp could hardly cherish such emotions. The fact is that the flag of no existing national state dare claim to be the symbol of the world-comradeship of humanity. It was with profound insight that the pioneers of the workers' movement chose the colour of blood for the banner of world community and solidarity, for the blood of animals and men is, after all, the internal medium which assures the co-operation of all the cells in that society of cells which is the body. And in the history of every people there have been inspiring times when they made their contribution to human progress. The British flag must remind us, not of the Factory Children and the Opium Wars, but rather of the Spirit of 1381, the New Model Army, the Floating Republic, and the Dorset Martyrs.

What a paradox that Haldane should have chosen *materialism* as the cause of all the trouble. Capitalist economics may indeed be inseparable from a mechanistic and atomistic sociology; the two arose together historically; Descartes and Boyle on the one hand, Petty and Gresham on the other. But the dialectics of nature, and dialectical materialism, are quite another matter.

The West and the East.

And so I come to the last paragraph of the last paper written by one of my principal teachers. "Materialism," he had written, "is a form of naïve belief which is easily understood by all nations, both

Western and Eastern, and it seems to me unfortunate that Western civilisation, with all the superficial advantages which the successful study of physical science confers upon it, has come to stand in the eyes of many in Eastern countries for little more than materialism. But materialism forms no basis for honesty, charity, regard for truth, loyalty, or art; and without these, real civilisation does not exist, and any apparent civilisation is quite unstable. The real strength of both Western and Eastern civilisations lies in their religions, and an understanding and respect for their religious ideas is essential for mutual understanding." No, it was not sufficient for the need of 1936. Between traditional metaphysical materialism and dialectical materialism no distinction was made. Yet to many of us it seems that precisely materialism is the best basis for honesty, charity, regard for truth, loyalty and art. Honesty hitherto may have been founded on philosophical idealism and the primacy of cognition or perception. But honesty has also been an almost impossible virtue in bourgeois civilisation, for the capitalist system of production is, and has always been, essentially predatory and atomistic, not co-operative. It was no coincidence that during the rise of capitalism at the Reformation and in the sixteenth century, there should have been so great a return, as represented in protestantism, to the Old Testament at the expense of the New. Honesty might play a minor part in the relations between merchants of the same town, but there were always Philistines or Egyptians whose exploitation could hardly reasonably be considered as displeasing to God. So much for honesty. Charity has become a word of abuse, implying the erratic bounty of wealthy robbers towards the poor and simple-hearted. Never will charity in its evangelical sense of love become a reality until the spiritual wickedness of class-distinctions is banished from the world. Regard for truth is in our civilisation confined to a few scientists and philosophers; it is not to be found in the realms of the patent medicine traffic, competitive advertising, unscrupulous journalism, artistically false nitwit entertainment and other like manifestations of the profit-making system. Loyalty of course is possible anywhere, but its value surely stands in proportion to the ideals which it serves. There is a loyalty of criminals one towards another; there is the loyalty of men and women towards their national flag, and at the other end of the scale there is the loyalty of often isolated individuals to the future *Civitas Dei*, the worldwide union of socialist republics, in which national sovereignty and human exploitation will seem as remote as the sacrifices of the Aztecs seem to us now. And lastly art, from being the product of the few for the few, will become the production of the many for the enjoyment of all.

But as regards what the East accepts from the West, and vice versa, Haldane was undoubtedly right. Mechanical materialism, plus idealistic religion, plus science for science's sake, plus capitalist economics, plus fascist armaments, class-stabilisation, and war-philosophy, will only spread misery, disappointment, and destruction. But there is a hope elsewhere, a red morning star. In playing my part at the funeral games of our lost kindly champion, what could I do else but point towards it?

And here we approach a remarkable historical paradox, a paradox of the greatest interest, but of which most people are still quite ignorant. The red morning star of progressive social philosophy, culminating in socialism and communism, has not always stood over Europe. In former times, it shone in the East, in more senses than one. Haldane discussed what the East was accepting from the West; it never occurred to him that the West owed something to the East. But there is a great debt, in general altogether unrealised, and here a brief explanation of it must be given.[1] I leave on one side, of course, the obvious debt of christian civilisation to Hebrew culture and what was born from it.

In traditional Western thought, the conception of "original sin" was dominant; Pelagius was the exception and the followers of Augustine were orthodox. It seems that the social consequences of this have never been fully explored, but it is extremely likely that a doctrine of original sin was of no small help to the property-owning classes in the propagation of the belief that the working masses could never hope to run a State organisation successfully. However one phrases it, the association between original sin and a pessimistic view of the possibilities of social organisation is unmistakable. Conversely, that optimistic view of the possibilities of social man which arose at the Enlightenment prior to the French Revolution in the eighteenth century, and which lies at the bottom of all subsequent optimistic social thought, was connected with a denial, tacit or avowed, of the conception of original sin. If the spirit of man is fundamentally good, then obviously the prospects of social justice are better than if the reverse is true. It may be granted that in strictly orthodox christian theology, the intrinsic goodness of the human soul had always been

[1] See the forthcoming book by Dr. E. R. Hughes, *The Great Learning in Action*.

maintained, with only a certain proneness to evil; but we are concerned here rather with the effect on the climate of thought than with the letter of the law on which a defence of christian theology might rely. The Encyclopaedists, therefore, were striking out a new line in their insistence on the intrinsic goodness of man's nature.

It now appears that in this progressive effort they were mightily assisted by a knowledge of the fact, then just entering Europe, that in traditional Chinese thought, the essential goodness of man's nature had always been a basic belief. In China Mêng-tze (Mencius) had been orthodox, and the pessimistic philosopher, Hsün-tze, had been heretical. The great story of the influence of China on European thought in the eighteenth century is only now coming to be written. An outward and visible manifestation of it was, of course, the Chinoiserie period which set its mark so thoroughly on all the arts of domestic decoration. But the philosophers of the Enlightenment were amazed, and deeply encouraged, to find that their belief in the separability of morality from superstitious religion, was and had been for more than a thousand years an essential tenet of classical Chinese philosophy.

About the beginning of the seventeenth century, the Jesuit missionaries reached Peking. They were highly cultivated men, who found the Chinese intellectual atmosphere so congruent with christian ideas that they felt they had but to add certain keystones to an arch already complete. Hence their very adequate translations of the Chinese Classics of Confucian philosophy; for example the *Confucius Sinarum Philosophus* of 1687, which includes Latin versions of the *Lun-Yü* and the *Ta-Shioh*. Such books contained notions of natural law, the possession of certain inalienable rights by common humanity, and above all, the view, enshrined to this day in the first schoolbook learned by every Chinese child (the three-character classic, the *San-Tze-Ching*), that man's spirit is fundamentally good. From this it was not a far cry to the slogan of Rousseau, "Man is everywhere born free, but he is everywhere in chains."

Evidence of the interest taken in classical Chinese philosophy by the Encyclopaedists, Rousseau, Diderot, Voltaire, etc. is easy to find. In Voltaire's *Dictionnaire Philosophique* one finds:—

"I knew a philosopher who had no portrait but that of Confucius in his working room, and underneath it he had written these lines:

'De la seule raison salutaire interprète,
Sans éblouir le monde clairant les esprits,
Il ne parla qu'en Sage, et jamais en Prophète;
Cependant on le crut, et même en son pays.'

"I have read his books with attention, and I have made extracts from them; I found nothing in them except the purest morality, without the least admixture of charlatanism."[1]

And in Germany, where the same movement was in progress, Christian Wolff delivered an inaugural address "Rede über die Sittenlehre der Sineser" in 1721 when appointed Pro-rector of the University of Halle. For this address, in which it was asserted that morality is independent of revelation, Wolff was so persecuted that he had to leave Prussia for twenty years.

We have, therefore, the remarkable paradox that China gave to the West a single precious treasure of doctrine, an idea that may yet save the West from itself, and the world from the domination of that Western commercialism against which Haldane wrote so strongly. Haldane did not wish the East to accept from the West mechanical materialism. In this he was right, though he did not couple with it what goes with it, other-worldly religiosity, servile science, and capitalist economics. He did not know that the West accepted from the East two centuries ago the germ of all those progressive social conceptions which alone can turn the spread of western civilisation throughout the world into beneficent channels. If christian theology is one grandmother of communism, confucian philosophy is the other.

---

[1] *Dictionnaire Philosophique*, 1834 edn. Vol. II, p. 406.

Limiting Factors in the History of Science, as Observed in the History of Embryology

(Carmalt Lecture at Yale University, 1935)

PROBABLY the best way to summarise the influences which have operated in the history of embryology is to concentrate attention on what may be called, borrowing a phrase from general physiology, the "limiting factors" of advance. We may thus regard the progress of knowledge about generation as governed by a reaction-chain, one link of which may at any given time be slower than all the others, and hence may set the speed for the whole.

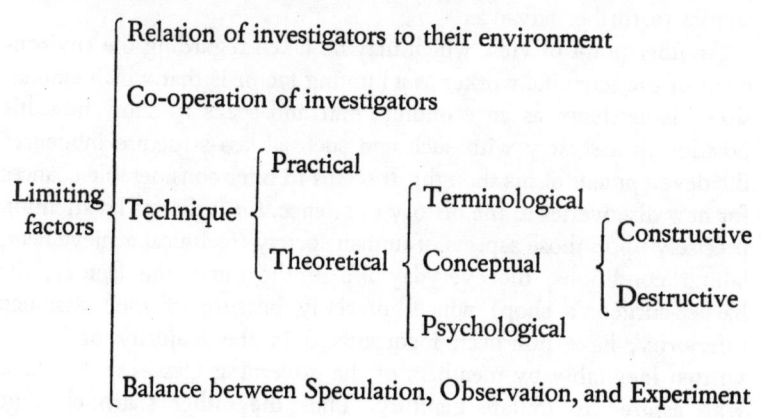

Limiting factors
- Relation of investigators to their environment
- Co-operation of investigators
- Technique
  - Practical
  - Theoretical
    - Terminological
    - Conceptual
      - Constructive
      - Destructive
    - Psychological
- Balance between Speculation, Observation, and Experiment

Scientists and their Environment.

Of these limiting factors the first which may be mentioned is the relation of investigators to their environment. The Carlylean tendency to regard the history of science as a succession of inexplicable geniuses arbitrarily bestowing knowledge upon mankind has now been generally given up as quite mythological. A scientific worker is necessarily the child of his time and the inheritor of the thought of many generations. But the study of his environment and its conditioning power may be carried on from more than one point of view.

A sharp distinction is made by the culture-historians, between the mental atmosphere of the Renaissance, the Baroque, the Rococo, the "Aufklärung" period, and so on. The political absolutism of the Baroque period mirrored itself in the extreme rationalism of seventeenth century biology. The Rococo period then brought in new movements towards freedom in the political sphere, and this took the form in science of a return to empiricism, so that the biological observations of Redi[1] and Wolff[2] were as much connected with the romantic movement as were the philosophical speculations of Rousseau. In the Encyclopaedists, the connection between empiricism in science and political freedom is particularly well seen. Men of learning examined the traditions of the technical arts and trades with new interests. The new eminence of the female sex in the Rococo period, unimaginable to previous ages, was perhaps connected with the temporary triumph of ovism.[3]

The social and political ruling ideas of an epoch thus play a large part in the scientific thought of the time, and may act as limiting factors to further advance.

Another point of view which may be taken regarding the environment of the scientific worker as a limiting factor is that which emphasises his existence as an economic unit and seeks to show how his position in a society with such and such a class-structure influences the development of his thought. It seems to offer considerable chances for new discoveries in the history of science, for it directs its attention precisely upon those aspects of human society (technical achievement, labour conditions, the everyday life of the mine, the factory, the barber-surgeon's shop) which, precisely because of their assumed inferiority, have not been incorporated in the majority of books, written inevitably by members of the governing classes or by those who aspired to imitate gentility. Thus the rather sharp cleavage between the philosophic biologist of the Hellenistic age, and the contemporary medical man, who might often be a slave, contributed to the sterility of ancient Mediterranean medicine, including obstetrics and gynaecology. In the later christian West there was not much

[1] *Esperienze ... alla generazione degl'Insetti*, 1668.
[2] *Theoria Generationis*, 1759.
[3] T. Bilikiewicz, *Die Embryologie im Zeitalter d. Barock u. Rokoko* (Leipzig, 1932). "Die Frau habe heute nicht nur das Recht, dass ihre Schönheit und Weiblichkeit in Dithyramben besungen werde; wenn sie den Platz auf dem Throne einnehmen oder über Throne verfügen könne, oder wenn sie im allgemeingesellschaftlichen Leben mit der wachsenden Gleichberechtigung immer verantwortlichere Rollen übernehmen könne, so habe sie auch das Recht, auf dem Gebiete der Embryologie dem männlichen Geschlechte in die Augen zu schauen als ein Wesen, das dieselben Rechte auf Freiheit habe. Der Ovismus liess diese Standarte wehen."

incentive to embryological study so long as the process of childbirth was left to the charms and incantations of barbarous midwives. But for a better insight into the economic position of embryologists in past ages nearly all the work remains to be done.

One necessity must constantly be kept before the mind's eye, namely, the knowledge of the relations between scientific thought and technical practice at any given period. For embryology this knowledge is difficult to acquire, since up to the time of the Renaissance obstetrics remained a part of primitive folk-medicine rather than of serious medical science. We see, however, in the publication of the Hellenistic gynaecological treatises in the 16th century (Bauhin,[1] Spach[2]) the satisfaction of a new demand, even though it took the typical Renaissance form of reprinting the Graeco-Roman classics. It was part of that movement to rationalise obstetrics which included William Harvey's *Exercitationes De Generatione Animalium*[3] and Malpighi's *De Formatione Pulli*,[4] and culminated in the celebrated man-midwives of the 18th century.[5] Again, the relation of the early systematists—Belon,[6] Rondelet,[7] Aldrovandus,[8] Ray[9]—to the beginnings of capitalist expansion is fairly clear, for the mediaeval bestiary could not cope with the influx of new animals and plants from hitherto unknown regions, any one of which might prove to be an exploitable commodity.

The Hellenistic divorce between scientific thought and empirical technique is an important case in point. Greek life was divided strictly into $\theta\epsilon\omega\rho\iota\alpha$ and $\pi\rho\hat{\alpha}\xi\iota\varsigma$; theory and practice. The latter was not thought fitting for a man of good birth. "Antiquity," says Diels,[10] "was entirely aristocratic in attitude. Even prominent artists, such as Pheidias, were classed as artisans, and were incapable of bursting through the barrier separating the workers and peasants from the upper class. A second cause of the slight technical progress in antiquity was its slave-holding system, which led to a lack of any impulse to develop the machine as a substitute for manual labour." Xenophon

---

[1] *Gynaeciorum*, etc., 1586. [2] Ibid., etc.," 1597.
[3] 1651, Eng. tr. 1653.   [4] 1672.
[5] E.g. The Chamberlens, Mauriceau, William Smellie, John Burton of York ("Dr. Slop") and Joseph Needham of Devizes; see the articles of Rosenthal, Janus, 1923, **27**, 117 and 192 and Mengert, Ann. Med. Hist., 1932, **4**, 453. The dissertation of Caspar Bose (*De obstetricum Erroribus*, 1729) is a typical attack on the midwives of the time.
[6] *Natural History of Fishes*, 1591, . . . *of Birds*, 1555.
[7] *De Piscibus Marinis*, 1554.   [8] *Ornithologia*, 1597.
[9] *Wisdom of God in Creation*, 1714.   [10] *Antike Technik* (Leipzig, 1920).

in the *Oeconomicus* held the industries in poor repute. "Men engaged in the mechanical arts," he says, "must ever be both bad friends and feeble defenders of their country." He troubled himself little with those skilful in carpentry, metallurgy, painting, and sculpture, but was always anxious to meet a "gentleman" (ὁ καλὸς καγαθός). The results of this were inevitable. Classical surgery and obstetrics benefited practically nothing from the speculations of the biologists from Alcmaeon to Herophilus. Surgeons and midwives remained members of the painter-cobbler-builder group, the group of base-born mechanics, entirely distinct from the astronomer-mathematician-metaphysician-biologist group, the group familiar with courts and tyrants.[1]

Only the greatest broke away from this tradition: Aristotle, when he conversed with fishermen; Archimedes perhaps, when he constructed his mechanical devices. For the rest, it was too strong. Down to the end of the Roman period the artillery in use remained precisely what it had been six hundred years before, although the empire was crumbling under barbarian pressure, and would have given anything, one would imagine, for an improved artillery capable of withstanding the Gothic armies. It is strange, as has been acutely said, that the Romans never invented anything so much in the Roman taste as a railway. So far as Hellenistic empirical industrial chemistry was concerned, the Democritean and Epicurean atoms might never have existed. And in medicine, the only effect of the brilliant Greek atomic speculations was to give rise to the Methodic school of Roman physicians, described by Allbutt,[2] the influence of which was never strong, and contributed relatively little to the main stream of therapeutics originating with Hippocrates.

In sum, we cannot dissociate scientific advances from the technical needs and processes of the time, and the economic structure in which all are embedded. We shall never understand the failure of Greek science if we consider it in abstraction from the environment which sterilised its speculation. The history of science is not a mere succession of inexplicable geniuses, direct Promethean ambassadors to man from heaven. Whether a given fact would have got itself discovered by some other person than the historical discoverer had he not lived

---

[1] In the Renaissance period Vesalius himself realised this: see B. Farrington, "Vesalius on the Ruin of Ancient Medicine," Modern Quarterly, 1938, **1**, 23. The question is also well discussed by J. G. Crowther in *The Social Relations of Science* (London, 1941), p. 378.

[2] *Greek Medicine in Rome* (London, 1921).

it is certainly profitless and probably meaningless to enquire. But scientific men, as Bukharin[1] said, do not live in a vacuum; on the contrary, the directions of their interest are ever conditioned by the structure of the world they live in. Further historical research will enable us to do for the great embryologists what has been well done by Hessen[2] for Isaac Newton, and in this survey it will not be out of place to take into account the social and economic status of the investigator himself (Cf. Frank Chambers[3] for the Hellenistic artist; M. Yearsley[4] for the sixteenth-century physician).

It would thus be of the greatest interest to know accurately the sources of the emoluments of embryologists at different times.[5] From Ornstein's admirable book on the scientific societies of the Renaissance,[6] the suspicion arises that their royal patronage was dictated not so much by a purely disinterested passion for abstract truth, as by the desire to profit as much as possible by the new techniques which the decay of the anti-usury doctrines, the willingness of the rising capitalist class to make industrial "ventures," and the far-ranging thought of the scientific men were combining to produce. In our own Royal Society, indeed, the preoccupation of the early Fellows with the "improvement of trade and husbandry" is patent to anyone acquainted with its early history (Cf. Thomas Sprat's account of it[7]). Thus Dr. Jasper Needham, elected in 1663, read only one paper before the Society—not, as might have been expected from his profession, on the transfusion of blood or the anatomy of the brain, but on the value and use of China Varnish.[8] However, it is probable that for the most part the embryologists whose work we shall have to discuss were practising physicians, free or relatively free, from the ancient tradition, and conscious that to understand the mystery of generation would be to advance the science and art of medicine.

In this connection, it is of interest that the Church in the 17th and 18th centuries provided a certain source of demand for embryological research. Of this Swammerdam[9] and Malebranche[10] provide

[1] *Historical Materialism* (Allen & Unwin, London, 1928).
[2] "The Social and Economic Roots of Newton's Principia" in *Science at the Cross-Roads* (London, 1931).
[3] *Cycles of Taste* (Harvard, 1928).
[4] *Doctors in Elizabethan Drama* (London, 1933).
[5] On this, cf. Cumston, Ann. Med. Hist., 1921, **2**, 265 and Dittrick, Ann. Med. Hist., 1928, **10**, 90.
[6] *The Role of Scientific Societies in the Seventeenth Century* (Chicago, 1928).
[7] *History of the Royal Society*, 1670.   [8] i.e. lacquer.
[9] *Biblia Naturae*, 1737,   [10] *Recherche de la Verité*, 1672.

interesting examples, and the conviction, then widely held, that research into the nature of generation would throw light on orthodox theological doctrines, such as that of original sin, led to an economic situation of value for biological development. Finally, it would be rash to minimise the factor of pure curiosity in seventeenth-century science. This was extremely marked in Leeuwenhoek's microscopical investigations, as Becking[1] has pointed out.

Scientific Co-operation.

Next comes Co-operation of Scholars. In the civilisation of the Hellenistic age, it may be said, a considerable measure of such co-operation had been attained; the works of Aristotle and Hippocrates were fairly readily available in written form, and evidence has been brought forward, particularly with regard to Jewish thought, that this was well used.

During the period when the biological school of Alexandria was at its height, that city became an important Jewish centre. Two centuries later it was to produce Philo, but now the Alexandrian Jews were writing that part of the modern Bible known as the Wisdom Literature. In books such as the Wisdom of Solomon, Ecclesiasticus, Proverbs, etc., the typical Hellenic exclusion of the action of gods in natural phenomena is clearly to be seen. There are two passages of embryological importance. First, in the book of Job (x. 10), Job is made to say, "Remember, I beseech thee, that thou hast fashioned me as clay; and wilt thou bring me into the dust again? Hast thou not poured me out like milk, and curdled me like cheese? Thou hast clothed me with skin and flesh, and knit me together with bones and sinews." This comparison of embryogeny with the making of cheese is interesting in view of the fact that precisely the same comparison occurs in Aristotle's book *On the Generation of Animals*. Still more extraordinary, the only other embryological reference in the Wisdom Literature, which occurs in the Wisdom of Solomon (vii. 2), also copies an Aristotelian theory, namely, that the embryo is formed from (menstrual) blood.

The Talmudists thought, moreover, that the bones and tendons, the nails, the marrow in the head and the white of the eye, were derived from the father, "who sows the white," but the skin, flesh, blood, hair, and the dark part of the eye from the mother "who sows the red." This is evidently in direct descent from Aristotle through

[1] Sci. Monthly, 1924, **18**, 547.

Galen, and may be compared with the following passage[1] from the latter writer's *Commentary on Hippocrates*: "We teach that some parts of the body are formed from the semen and the flesh alone from blood. But because the amount of semen which is injected into the uterus is small, growth and increment must come for the most part from the blood." It might thus appear that, just as the Jews of Alexandria were reading Aristotle in the third century B.C. and incorporating him into the Wisdom Literature, so those of the third century A.D. were reading Galen and incorporating him into the Talmud.

But we must beware here of suffering a distortion of perspective in the contemplation of antiquity, for it is easy to exaggerate the co-operation of ancient thought. A single idea could consider itself lucky if it passed once in twenty-five years between Greece and India after Alexander. Among the conflicting influences that gave rise to the civilisation of the later West, this co-operation, hampered by enormous linguistic difficulties on the one hand and by the diversion of interest from scientific to ethical and theological channels on the other, sank to a very low level. Hence we have the remarkable spectacle of a Leonardo, many years ahead of his contemporaries, and able to earn a living only as a designer of fortifications, finding it impossible to communicate his discoveries to any living person, and reduced to burying them in note-books,[2] only by a mere chance available to scholars of after ages.

Scientific Technique.

Among the most important of limiting factors we must reckon Technique, extending the term to cover mental as well as material methodology. The part which the latter has played in the history of embryology can hardly be overrated. Thus until the introduction of hardening agents, especially alcohol, by Boyle,[3] the examination of the early stages of embryos was bound to remain crude, and embryology attained an entirely different level immediately afterwards, in the hands of Maître-Jan.[4] The parallel case of the microscope is too familiar to dwell on, but the work of Malpighi obviously marked a

[1] Ten centuries later it was still worth while for Harvey to have a hit at this opinion. "In the interim," he says (1653, p. 116), "we canot chuse but smile at that fond and fictitious Division of the Parts into Spermatical and Sanguineous; as if any part were immediately framed of the semen, and were not all of one extract and original."

[2] *Quaderni d'Anatomia*, ed. Vangensten, Fohnahn & Hopstock (Copenhagen, 1911).

[3] Phil. Trans. Roy. Soc., 1666, **1**, 199.

[4] *Observations sur la formation du Poulet, etc.*, 1722.

turning-point in the science. It may here be noted, however, that even when methods are available, the workers of the time do not necessarily use them, and although Harvey could have employed an early form of microscope, he voluntarily restricted himself to the weak lenses, "perspicilia," or perspectives, which had already been used by Riolanus. A still more obvious instance is that of artificial incubation. Carried on in Egypt since the remotest antiquity, this process must have been at the disposal of Egyptian physicians, Alexandrian biologists, and Arabian scholars for a period of three thousand years, yet so far as we know, no embryological use of it was ever made. In eighteenth-century France and England the technique of the process had to be painfully rediscovered at a time when biologists were only too eager to make use of such assistance. Let us mention, as other instances of the effect of material technique on embryology, the burst of knowledge which followed the invention of the automatic microtome by Threlfall[1] and others about 1860, and the great advance which in our own century has followed the successful mastery of grafting technique by Spemann.[2]

Just as important, however, as material technique is mental technique. And first with respect to words; on several occasions we have had to notice a standstill on account of the lack of a satisfactory terminology. Thus in the thirteenth century Albertus of Cologne[3] had arrived at a point beyond which progress was impossible in the absence of new words. When, for example, there was no other means of describing the sero-amniotic junction in the hen's egg than by speaking of "the hole on the left side of the vessel which runs above the membrane on the right hand of something else," accuracy was difficult and speed impossible. A precisely similar position was occupied by Boerhaave[4] in the eighteenth century, only now in the case of biochemical words. Faced with some substance such as a "greasy, streaky yellow oil, smelling of alkaline salt" Boerhaave was unable to describe it except in these common-sense terms, and lacking the means either to submit it to further analysis or to characterise it by accurate physico-chemical constants, he was forced to admit a vast number of ultimates into his schemes which were not ultimate at all.

Mental technique, as a limiting factor in embryological history,

---

[1] See Biol. Rev., 1930, 5, 357.
[2] See my *Biochemistry and Morphogenesis* (Cambridge, 1942).
[3] *De Animalibus*, ed. Stadler, 1916.   [4] *Elementa Chemiae*, 1732.

goes deeper than words, however, for it involves the concepts of the investigator. What the Germans call "Begriffsbildung" or the construction of concepts congruent with certain sorts of natural phenomena, though never conscious in the history of biology, has none the less been operative. In this field we may remember the doctrine of Galen concerning the natural faculties ($\delta\upsilon\nu\acute{\alpha}\mu\epsilon\iota\varsigma$), and the immense length of time which was required for biologists to see that it was nothing more than a concise statement of the phenomena themselves. Not until it was "seen through" as an explanation was post-Renaissance biology possible. Similarly, the peculiar contribution of Leonardo to embryology was his realisation that embryos could be measured at a succession of moments. The application of the concept of change in weight and size with time, a concept which, as modern biology shows, admits of much accuracy when properly worked out, was thus first made by Leonardo. In the same way Boyle[1] was the first to see clearly that a problem of *mixture* is presented by the developing embryo (though Hippocrates had stated it dimly some two thousand years before). If the embryo is made up of mixed things, some definite proportion and way of mixture must exist. And no hope of finding out what this was could be obtained from the Aristotelian elements (heat, cold, moisture, and dryness) or from the Alchemical principles (salt, sulphur, and mercury). Hence Boyle's emphasis on the corpuscularian or mechanical hypothesis, and all its historical implications.

His preference for the "mechanical or corpuscularian" philosophy was mainly due to his realisation that unless chemistry was going to start measuring something it might as well languish in the obscurity to which Harvey would willingly have relegated it. Thus he says:—

> "But I should perchance forgive the Hypothesis I have been examining (that of the alchemists), if, though it reaches but to a very little part of the world, it did at least give us a satisfactory account of those things which 'tis said to teach. But I find not that it gives us any other than a very imperfect information even about mixt bodies themselves; for how will the knowledge of the *Tria Prima*[2] discover to us the reason why the Loadstone drawes a Needle, and disposes it to respect the Poles, and yet seldom *precisely* points at them? How will this hypothesis teach

[1] *The Sceptical Chymist*, 1661.
[2] The three alchemical (hypostatical) principles.

us how a Chick is formed in the Egge, or how the seminal principles of mint, pompions, and other vegetables, can fashion Water into various plants, each of them endow'd with its peculiar and determinate shape and with diverse specifick and discriminating Qualities? How does this hypothesis shew us, *how much* Salt, *how much* Sulphur, *how much* Mercury must be taken to make a Chick or Pompion? and if we know that, what principle is it, that manages these ingredients and contrives, for instance, such liquors as the White and Yolke of an Egge in such a variety of textures as is requisite to fashion the Bones, Arteries, Veines, Nerves, Tendons, Feathers, Blood and other parts of a Chick; and not only to fashion each Limbe, but to connect them altogether, after that manner which is most congruous to the perfection of the Animal which is to consist of them? For to say that some more fine and subtile part of either or all the Hypostatical Principles is the Director in all the business and the Architect of all this elaborate structure, is to give one occasion to demand again, *what proportion and way of mixture* of the *Tria Prima* afforded this Architectonick Spirit, and what Agent made so skilful and happy a mixture?"

Boyle's instance of the magnetic needle pointing nearly, not exactly, at the north, and his use of the expressions "how much," "how many," "proportion," "way of mixture," indicate that he was moving towards a quantitative chemistry, and by express implication a quantitative embryology. Elsewhere he says that he thinks the *Tria Prima* will hardly explain a tenth part of the phenomena which the "Leucippian" or atomistic hypothesis is competent to deal with. Thus, although Boyle made few experiments or observations on embryos, he occupies a very important position in the history of embryology.

Allied to this creation of concepts, and the choice of one of them to apply, we find that the mentality of the workers of the past has often been particularly different with regard to a quality which can only be called Audacity. Probably Aristotle's greatest claim to our respect is that alone of his contemporaries and predecessors he had the audacity to suggest that animal form was not limitlessly manifold or infinite in its manifestations, but that given industry and intelligence, a classification was possible. This alone marks him out above all subsequent biologists. On a smaller scale, we find the same mental audacity in Kenelm Digby,[1] whose discussions of the development of the chick are remarkable for their naturalistic tone, for their conviction that the processes of development are not beyond the reach of the reason and imagination of man. It is ironic that Digby, who did little or nothing himself to advance our knowledge, should have spoken thus, and that his great contemporary, William Harvey, to whom we are indebted for so many advances in embryology, was led to despair of understanding development. Another interesting point that emerges from the same period is that such mental audacity can go, perhaps, too far, as when Descartes[2] and Gassendi[3] built up an embryology *more geometrico demonstrata*, in which the facts were relegated to an inferior position and the theory was all.

But not only must the right concepts be chosen, the wrong ones must be abandoned. One of the principal necessities which has faced investigators since the earliest times has been the recognition of silly questions in order to leave time for the examination of serious ones. It was presumably inevitable that the pseudo-problems concerning the entry of the soul into the embryo should be taken seriously until a very late date. But a more typical instance of a meaningless question may be found in the dispute about what parts of the egg *form* the chick and which *feed* it. The tacit assumption here was that since to common sense food and flesh are different things, there must be in the hen's egg, aside from a sufficient provision of food, some sort of pre-flesh out of which the embryo can be made. Not until 1651 did this pseudo-problem go out of currency in the light of Harvey's demonstration of the unsoundness of the assumption.

The expulsion of ethics from biology and embryology forms another excellent example. That good and bad, noble and ignoble, beautiful and ugly, honourable and dishonourable, are not terms with a biological meaning is a proposition which it has taken many centuries for biologists to realise.

Ideas of good and bad entered biology partly under the concept of "perfection." In 1260 Albertus was maintaining that male chicks always hatched from the more spherical eggs and female chicks from the more oval eggs, because the sphere is the most perfect of all figures in solid geometry, and the male the more perfect of the

---

[1] *Two Treatises, in the one of which the Nature of Bodies, in the other the Nature of Man's Soule, is look'd into, in way of discovery of the Immortality of Reasonable Soules*, 1644.

[2] *L'Homme, et la formation du Foetus*, etc., 1664.   [3] *Opera*, 1658.

two sexes. We realise to-day that to ask which is the more perfect of the two sexes is a meaningless question, for we have expelled ethics from science and cannot regard any one thing as being more perfect than any other. Again, describing the course of the arteries in the developing chick, Albertus says: "One of the two passages which springs from the heart branches into two, one of them going to the spiritual part which contains the heart, and carrying to it the pulse and subtle food from which the lungs and other spiritual parts are formed; and the other passing through the diaphragm to enclose the yolk of the egg, around which it forms the liver and stomach." This distinction between the organs above the diaphragm—the lungs, heart, thymus, etc.—called "spiritualia," and the organs below—the stomach, liver, intestines, spleen, etc.—runs through the whole of the early anatomy. It was as if the organs of the thorax were regarded as a respectable family living at the top of an otherwise disreputable block of flats. To us it seems absurd to call one organ more "spiritual" than another, but that is because we realise the irrelevance of ethical issues in biology. Thomas Aquinas, about the same time, in his *Summa Theologica*, dealt in passing with human generation. "The generative power of the female," he said, "is imperfect compared to that of male, for just as in the crafts, the inferior workman prepares the material and the more skilled operator shapes it, so likewise the female generative virtue provides the substance but the active male virtue makes it into the finished product." This is really the pure Aristotelian doctrine, but St. Thomas gives it the characteristically mediaeval twist. Aristotle might make a distinction between form and matter in generation, but the feudal mentality, with its perpetual hankering after status, would at once enquire which of the two, male or female, was the higher, the nobler, the more honourable.

In the eighteenth century the same frame of mind persisted. It was maintained that in every detail of the visible world some evidence could be found for the central dogma of natural theology, the belief in a just and beneficent God.[1] Between 1700 and 1850 a multitude of books were written which purported to reveal the wisdom and goodness of God in the natural creation. The theologians took what suited their purpose and left the rest. It is instructive to see how Goethe, who was somewhat committed to the theological interpretation of phenomena, reacted to the ornithological anecdotes of

[1] For a striking example of this, see Edmund Gosse's *Father and Son* (London, 1907, and many later editions).

his secretary Eckermann on October 18, 1827.[1] He said little while Eckermann told him about the habits of the cuckoo and other birds, but when Eckermann related how he had liberated a young wren near a robin's nest and how he had found it subsequently being fed by the robins, Goethe exclaimed: "That is one of the best ornithological stories I have ever heard. I drink success to you and your investigations. Whoever hears that, and does not believe in God, will not be aided by Moses and the prophets. That is what I call the omnipresence of the Deity, who has everywhere spread and implanted a portion of His endless love." And so it always was with the theological naturalists; they hailed with enthusiasm the discovery of monogamy in tortoises, or mother-love in goats, but they had nothing to say concerning the habits of the hookworm parasite or the appearance of embryonic monsters in man. Not until the beginning of the nineteenth century did it become clear that nature cannot be divided into the Edifying, which may with pleasure be published, and the Unedifying, which must be kept in obscurity.

The Balance of Thought and Observation.

In the end we may say that the progress of a branch of natural science such as embryology depends on a delicate balance of three things; speculative thought, accurate observation, and controlled experiment. Any modification of the optimum balance will act as a powerful limiting factor on progress. Speculative thought, in particular, has shown a tendency to crystallise too readily into doctrines which, by way of attachment to some philosophical or theological issue, live a longer life than they deserve. Thus the Aristotelian theory of the formation of the embryo by the coagulation of the menstrual blood, built in the first instance upon a faulty deduction, became incorporated in the Aristotelian tradition of *forma* and *materia*, and although quite repugnant to observation, remained the official theory throughout the European middle ages, and apparently in perpetuity in India. So powerful was the rationalism of a medical education at about 1630 that the physicians to whom Harvey demonstrated the empty uteri of the King's does preferred to believe their books rather than the evidence of their senses.

The account given by Harvey himself (1653, p. 416) cannot be omitted:

[1] *Conversations with Goethe*, Everyman edition, p. 243.

"When I had often discovered to His Majesties sight this alteration in the Womb, and having likewise plainly shewed that all this while no portion of seed or conception either was to be found in the Womb, and when the King himself had communicated the same as a very wonderful thing to diverse of his followers, a great debate at length arose: The Keepers and Huntsmen concluded, first, that this did imply, that their conception would be late that year, and thereupon accused the drought; but afterwards when they understood that the rutting time was past and gone; and that I stood stiffly upon that, they peremptorily did affirm, that I was first mistaken my selfe, and so had drawn the King into my error; and that it could not possibly be, but that something at least of the Conception must needs appear in the Uterus: untill at last, being confuted by their own eyes, they sate down in a gaze and gave it over for granted. But all the King's Physicians persisted stiffly, that it could no waies be, that a conception should go forward unless the males seed did remain in the womb, and that there should be nothing at all residing in the Uterus after a fruitfull and effectuall Coition; this they ranked amongst their ἀδύνατα.

Now that this experiment which is of so great concern might appear the more evident to posterity; His Majestie for tryal-sake (because they have all the same time and manner of conception) did at the beginning of October separate about a dozen Does from the society of the Buck and lock them up in the Course near Hampton Court. Now lest any one might affirm that doubtlessly there did continue the seed bestowed upon them in Coition (their time of Rutting being then not past) I dissected divers of them, and discovered no seed at all residing in their Uterus; and yet those whom I dissected not, did conceive by the virtue of their former Coition (as by Contagion) and did Fawn at their appointed time."

And precisely parallel to this attitude was that of the preformationists in the following century, who, having decided, like Bonnet,[1] that epigenesis was inconceivable, only accepted such observations as confirmed their *a priori* view.

Preformationism as a manifestation of rationalism merits further examination. The dogmatic manner in which preformationism was

---

[1] *Considerations sur les corps organisées*, 1762.

held during the eighteenth century would not perhaps have been so crushing if the biologists of that time had been able to take a mathematical argument more seriously. There was Harvey's very convincing argument about the circulation of the blood, and Freind's[1] equally convincing (though unfortunately in tendency erroneous), argument about the quantity of menstrual blood and the weight of the newborn foetus. Verbally, it was still quite possible to support the Hellenistic view that the embryo was formed from menstrual blood, in the post-Harveian period, if it were admitted that this blood flowed little by little through the umbilical vessels. This was the position of John Freind in his treatise on menstruation, *Emmenologia* (1700-30). Calculating the amount evacuated in nine months, he said: "The quantity of Blood which the Mother may bestow upon the nourishment of her Offspring will be *lib.* 13 ʒ 2⅓, which will outweigh the newborn Foetus with all its Integuments, if they should be put into a Balance; and leave no room to doubt, its being able to bestow very proper nourishment on the Embrio. For the mean weight of a new-born Foetus is about *lib.* 12, some-times it is found greater, and very often less."

If these could have been accepted, it was a pity that Hartsoeker's argument about preformation could not. In 1722 Hartsoeker[2] calculated that $10^{100,000}$ rabbits must have existed in the first rabbit, assuming that the creation took place 6,000 years ago and that rabbits begin to reproduce their kind at the age of six months. But to this Bonnet merely answered that it was always possible, by adding zeros to units, to crush the imagination under the weight of numbers, and he described the performation theory as one of the most striking victories of the understanding over the senses. It would have been better described as one of the most striking victories of the imagination over the understanding.

The fact is that the biologists of the eighteenth century, carried away by preformationist theory, took embryology on to a plane where observation became superfluous. They would have found acceptable the sentiment satirised by Boyle that "it is much more high and philosophical to argue *a priori* than *a posteriori*," and were eventually debarred from looking at developing embryos by their conviction that structure and organisation would certainly be there, whether they could see it or not. The preformationist controversy was, in fact, a repetition in biology of the controversy between the

---

[1] *Emmenologia*, 1720.   [2] *Receuil de plusieurs pièces de physique*, 1722.

rationalists and the empiricists in philosophy. The contemporary rationalists were people who held that "human beings were in possession of certain principles of interpretation which were not simply generalisations from experience, but could nevertheless be used as major premises in arguments concerning nature. If observations were not in accordance with expectations founded on such reasoning, they were dismissed as illusions. The empiricists, on the other hand, held that there was no knowledge independent of observation, and that the rationalists' principles, in so far as they were admissible at all, *were* generalisations from experience." It is obvious that nearly all the preformationists were rationalists. They thought that Reason was in a position to decide the issue whatever might be the results of observation. "It is remarkable," as Cole says,[1] in his book on this period, "that the preformationists did not realise that if the point to be established is assumed at the outset all further discussion is superfluous." In this example, then, we have a disturbance of the balance towards the side of rationalistic speculation.

It would be a mistake, however, to regard this tendency as confined to the eighteenth century. Ample examples of its presence can be collected from nearly every period in biological history. "We plume ourselves," says Cole, "on that aspect of our work which is vain and argumentative, and condescend to the more modest but enduring labour of observation." There can be no doubt that this state of affairs, so unfortunate for science, is one aspect of that contempt for manual labour which has run through the stratified structures of all societies in the history of civilisation. The manipulator of paper and ink, educated in the classical traditions of his time, has always seemed, by reason of his superficial similarity to the political administrator, a superior being to the empirical mechanic engaged in the manual work of the arts and industries. The tradition is as old as civilisation, yet for the advance of science it must be broken. Not until the manual worker and the audacious theorist are combined in one person will the fullest development of scientific thought be possible.

All the greatest experimental scientists are evidence of this, but by no means all of them have been conscious of it. It is therefore of particular interest to read the words which the great Russian physiologist, Pavlov, wrote to a meeting of Stakhanovist miners, in 1935.

[1] *Early Theories of Sexual Generation* (Oxford, 1930).

"All my life," he said, "I have loved and still love both intellectual and manual work, and the second perhaps even more than the first. Especially have I felt satisfaction when into the latter I have been able to carry some good problem, thus uniting head and hands. Upon this same road you are travelling. With all my heart I wish that you may advance far along this path, the only path that secures happiness for man."[1]

Experiment in Embryology.

When I once gave some lectures on this subject at University College, London, they bore the title "Speculation, Observation and Experiment as Illustrated by the History of Embryology." Of the first two of these factors we have seen enough, but the third would necessitate the continuation of the story down to the end of the nineteenth century. The true science of experimental embryology did not come into being until the time of Wilhelm Roux.[2] The early chemical observations on the embryonic liquors were indeed observations rather than experiments; and there was no systematic study of the changes which the liquors undergo during the development of the foetus; this was not done till the time of John Dzondi.[3] Harvey's segregation of does at Hampton Court merits, no doubt, the name of experiment, involving as it did, the use of "controls," and an outstanding instance is the ligature of Nuck[4] in 1691. The work of Nuck is very important, as one of the earliest instances of experimental procedure. He ligatured the uterine horns after copulation in a dog, and observed pregnancy afterwards, implantation having taken place above the ligature. His conclusion was that the embryo was derived from the ovary and not from the sperm—"*animal ex ovo generari experimento probatur.*"

As in Nuck's case, experiment in the hands of both Spallanzani[5] and J. T. Needham[6] led to error. Spallanzani confuted his adversary on the question of spontaneous generation and the vegetative force by what amounted to rigid criticism of experimental conditions, but later on denied their proper function to the spermatozoa on exactly the same methodologically faulty grounds.

---

[1] Quoted in *Lectures on Conditioned Reflexes*, vol. II, p. 53 (London, 1941).
[2] *Gesammelte Abhandlungen ü. Entwicklungsmechanik*, 1895.
[3] Journ. f. Chem. u. Physik, 1806, 2, 652.
[4] *Adenographia curiosa*, 1691. [5] *Saggi di osservazioni microscopiche*, etc., 1766.
[6] *New Microscopical Discoveries*, 1745.

Nevertheless, experimentation, the active interference with the course of nature and the subsequent observation of the resulting system in comparison with systems in which no such interference has taken place, was a characteristically nineteenth-century product as far as biology and embryology are concerned. Only at the present day, indeed, are we beginning to appreciate the statistical and other difficulties attending upon the full application of the experimental method to living organisms, and the manifold obstacles which prevent obedience to the rule that only one variable be modified at one time. But this is no matter of reproach against the older embryologists. Knowledge of form must necessarily precede knowledge of change of form and the factors producing it, and so we see during the last seventy years the production of "Normaltafeln" or tables of morphological pictures showing normal development; these are the essential basis for experimental studies.

On the other hand, there can be no doubt that a plethora of observation and experiment is also bad for scientific progress. Modern biology is the crowning instance of this fact. What has been well called a "medley of *ad hoc* hypotheses" is all that we have to show as the theoretical background of a vast and constantly increasing mass of observations and experiments. Embryology in particular has been theoretically backward since the decay of the evolution theory as a mode of explanation. Embryologists of the school of F. M. Balfour[1] thought that their task was accomplished when they had traced a maximum number of evolutionary analogies in the development of an animal. Wilhelm His,[2] perhaps the first causal embryologist, struggled successfully to end this state of affairs. "My own attempts," he wrote in 1888 in a famous passage,[3] "to introduce some elementary physiological or mechanical explanations into embryology have not been generally agreed to by morphologists. To one it seemed ridiculous to speak of the elasticity of the germinal layers: another thought that by such considerations we put the cart before the horse; and one recent author states that we have something better to do in embryology than to discuss tensions of germinal layers, etc., since all embryological explanation must necessarily be of a phylogenetic nature." But this strictly evolutionary dominance in embryology did not last on into the twentieth century. The unfortunate thing is that nothing has so far been devised to put in its place. Experimental embryology,

[1] *Comparative Embryology*, 1880.
[2] *Unsere Körperform*.
[3] Proc. Roy. Soc. Ed., 1888, **15**, 294.

Morphological embryology, Physiological embryology, and Chemical embryology form to-day a vast range of factual knowledge, without any great unifying hypothesis, for we cannot as yet dignify the axial gradient doctrines, the field theories, and the speculations on the genetic control of enzymes, with such a position. We cannot doubt that the most urgent need of modern embryology is a series of advances of a purely theoretical, even mathematico-logical, nature. Only by something of this kind can we redress the balance which has fallen over to observation and experiment; only by some such effort can we obtain a theoretical embryology suited in magnitude and spaciousness to the wealth of facts which contemporary investigators are accumulating day by day.

The Biological Basis of Sociology

(Based on a contribution to the Second Conference
on the Social Sciences, 1936)

THERE are two extreme views which might be taken concerning the significance of biological data and conclusions for sociological thought. It might be held, either that our present biological knowledge has no relation to the understanding of social phenomena; or conversely, that it is more important than any other illumination we can expect to have. The first of these views might perhaps be taken by some old-fashioned theologian, but probably no reader of this book would support it. The second has a more subtle appeal, and has been rather widely held. Nevertheless, I believe it to be exceedingly dangerous. I believe, on the contrary, that while our present biological knowledge can furnish us with many essential guiding principles and clues in the study and direction of social phenomena, it must at all points be supplemented by principles derived from observation of the social facts themselves and applying only to those facts. In the world of nature we have to deal with a succession of levels of complexity and organisation. The principles which apply to one of these levels do not apply to the others, although at every level the principles appropriate to the lower levels must be taken into account, modified though they may be by the special new conditions prevailing. At the level of life itself, this doctrine is not vitalism, any more than it is "crystallism" to demonstrate the special laws which govern the behaviour (to use a convenient *façon de parler*) of liquid and solid crystals.[1] Similarly, in the associations of highly cerebrated ape-like organisms which are studied by sociology, new principles apply, principles which are meaningless when mentioned in connection with lower levels, principles which may have manifestations familiar to us by quite other names, such as purpose, the good life, social cohesiveness, love, etc.[2]

[1] See the author's *Order and Life*, pp. 46, 47.
[2] After writing the above, I found the following:
   "The laws of social development are specific laws. It is, therefore, fundamentally incorrect and methodologically impermissible to transfer mechanically laws of a biological order into the province of social development" (Bukharin, N., in the Marx Memorial Volume of the Moscow Academy of Sciences, 1933, Eng. tr., *Marxism and Modern Thought*, p. 33).

Biology, therefore, can contribute greatly to sociological study and social control, but its word can never be the last one. Sociologists and social engineers may neglect it at their peril but equally they must master it and never let it master them.

The Denial of the Sociological Level; "Biologism."

It may be advisable to give a few illustrations of the melancholy consequences which follow abandonment of this philosophic principle. Recent articles by E. W. McBride will do. In a letter to Nature,[1] this writer (a professional biologist) opened the question of the cultivation of the unfit. Describing a certain physical deterioration which followed the isolation of chamois and red deer in special parks separated from their natural enemies, he went on to speak of the "elaborate and costly social services" which keep alive the "morally, mentally, and physically" human unfit. In his opinion, sterilisation to prevent further reproduction should be applied to all those who "have to resort to public assistance in order to support their children," although at the same time he was at pains to point out that sterilisation must be regarded as a trauma and hence as a punishment. "If the deep-seated sub-conscious desire to perpetuate himself by producing offspring is rendered impossible," he said, "however slight and painless the operation may be, it leaves a psychic wound which will never heal." Now the apparently guileless transition from the red deer to the unemployed workmen of a civilised country covers several *non sequiturs*. We are given no evidence that the deformed specimens of deer were mentally or morally unfit, or if so, unfit for what? One would have thought that childhood fairy stories apart, adult reasoning would not attribute mental and moral qualities to animals. Fitness or unfitness in men or animals, says McBride, is ability or inability to maintain themselves in their normal environments. But what is a normal environment? Do not men differ from animals precisely in that their environment is partly of their own creating? Do they not have the power to alter their environment and to make it anew, in nearer accordance with their wishes? We clearly see that endless difficulties are created by the refusal to admit that social phenomena are upon a different level from biological phenomena, and that thinkers who will not admit the difference are the most dangerous of guides. One must seek some further explanation of the

[1] E. W. McBride, Nature, 1935, **137**, 44, further correspondence, p. 188.

sadistic attitude of this writer towards the working class, other than the conclusions of a sound sociology.

Sterilisation, indeed, is perhaps the best example of the dangerous possibilities now at the disposal of the advocates of class and race oppression who desire to put their proposals on a scientific basis. J. B. S. Haldane[1] in his Lockyer Lecture for 1934 instanced the case of a labourer recommended for sterilisation in the state of Washington, U.S. after conviction for stealing food. The whole family was of sub-normal intelligence and had been for a long period on the verge of starvation. But as Haldane says it did not occur to the judge either that there might be any connection between the starvation of the children and their mental dullness, or that there was anything wrong with conditions under which a beet-sugar labourer could not earn enough to support five children. The review of Vignes[2] shows that there *are* cases where sterilisation is suitable. But Jennings,[3] in his acute enumeration of current popular biological fallacies, points out that many people think that if we prevent the breeding of hereditary defectives we shall largely get rid of such defectives in future generations, and conversely, that superior people come from superior parents, and that these things will continue to happen. Any society that carries out this plan will be in for a great disappointment. For unfortunately the great majority of defective genes which cause the troubles we want to get rid of are contained in normal people who carry them just as "typhoid-carriers" carry the bacilli of typhoid fever without ever getting it themselves. Thus 0.3 per cent of the world's population is actually suffering from feeblemindedness, but no less than 10 per cent is composed of people carrying the defective genes, while normal themselves. R. A. Fisher's calculations[4] showed that about 11 per cent of the feebleminded of any generation come from the mating of the feebleminded of the previous generation, while 89 per cent of them come from matings among the carrier group. The converse case, that of genius, almost certainly works in the same way, according to Raymond Pearl.[5] It is unfortunately only too

[1] J. B. S. Haldane, *Human Biology and Politics*, Lockyer Lecture, British Science Guild, 1934.
[2] H. Vignes, La Presse Medicale, 1934, May 19th and June 13th.
[3] H. S. Jennings, *The Biological Basis of Human Nature* (London, 1930).
[4] R. A. Fisher, Journ. Hered., 1927, **18**, 529. See also J. B. S. Haldane, *New Paths in Genetics* (Allen & Unwin, London, 1941), esp. p. 115 ff.
[5] R. Pearl, Amer. Mercury, 1927, **12**, 257. Pearl employed the following method Genius was, for the present purpose, defined as that which entitles its bearer to more than one page of space in the *Encyclopaedia Britannica*. Among the 126 parents of 63

obvious how easily a weapon such as sterilisation may be used indiscriminately by a fascist government at the dictates of racialist theories entirely without serious scientific foundation, as may be read in the book of Huxley, Haddon & Carr-Saunders.[1] If we may judge from accounts of "national-socialist biology" such as that of Brohmer,[2] such political sterilisation is an actual fact to-day.

I come back to the question of fitness and unfitness, and the relative parts played by inheritance and environment in moulding human material. The stupid acceptance of the conditions holding good in the animal world, and their application to human societies, did not originate with McBride. Already in the last century the conception of the struggle for existence in the world of animals introduced by Darwin under the name of natural selection, unfavourable or lethal variations being weeded out by the press of circumstances, was appealed to as the justification for *laissez-faire* economics. Such a book as that of Headley,[3] *Darwinism and Modern Socialism*, dating from 1903, well illustrates this trend, and the great embryologist, O. Hertwig, found it necessary to write a book expressly condemning the sociological application of the principle of natural selection.[4] For men of the type of Headley, socialism conflicted with ineradicable human instincts and could only thrive as a theory. In this way, the atomistic and individualistic character of the system of capitalist enterprise, which had already in the seventeenth century received support from the concurrent theories of physical atomism, was now to be based on the perpetual and universal strife for food and generation found in the world of wild animals. It was, of course, suggested, as by Engels[5] and later by Pannekoek,[6] that the theory of natural selection was the effect, not the cause, of capitalist production, since Darwin's mind cannot but have been receptive to the atmosphere of the society in which he found himself; and certainly it is undeniable that the

---

philosophers of genius, only three were themselves sufficiently distinguished to leave behind them any record of the fact. Among the 170 parents of 85 poets of genius, only three (again) were sufficiently distinguished for posterity to be aware of it.

[1] J. S. Huxley, A. C. Haddon & A. M. Carr-Saunders, *We Europeans* (London, 1935).

[2] P. Brohmer, *Mensch-Natur-Staat; Grundlinien einer nazional-sozialistischen Biologie* (Frankfurt, 1935).

[3] F. W. Headley, *Darwinism and Modern Socialism* (London, 1909).

[4] O. Hertwig, *Zur Abwehr des ethischen, sozialen, und politischen Darwinismus* (Jena, 1921).

[5] F. Engels, *Dialectics of Nature*, p. 641.

[6] A. Pannekoek, *Marxism and Darwinism* (Chicago, 1912).

theory originated from Malthus, since Darwin says so himself in his autobiography. We thus have a theory arising from an absurdly easy (as it now seems to us) acceptance of discomfort in human societies, applied with much success to a field where it more properly belonged, that of animals in natural surroundings, and then brought back again with all the resounding prestige which it acquired from its biological success, to justify sociological states of affairs to which it was perfectly inapplicable. In reading Malthus I find nothing more amazing than the mental blindness with which he approached the subject of birth control. "Improper Arts," as he called them, being utterly out of the question, the population *must* increase more rapidly than the available means of subsistence, and struggle must ensue. If capitalist competition and misery due to under-production could in those days be represented as unavoidable natural laws, there can certainly be no sense in so regarding them in an age when birth-control is rationally considered and when the advance of scientific technology has rendered under-production a thing of the past. It seems quite clear that a control of population by a world-authority subject to no class or race prejudices, is the goal towards which all our efforts should be tending.

The Mechanisation of Democracy.

Reference was made just now to the fallacy that most of the mental deficiency in a population arises from the reproduction of mentally deficient people. Allied to the above fallacy is that which alleges that biology requires an aristocratic constitution of society. Each grade of society, it is said, reproduces itself. Criminals produce criminals, public school tie men produce public school tie men, intellectuals produce intellectuals. The children of Plato's guardians become guardians too.[1] Whereas the fact is that from the higher many lower are produced, from the lower many higher, and from the great mediocre mass are produced more of the higher than the higher produces itself, and more of the lower than the lower produces itself. *A democracy that can produce experts* is emphatically the form of society called for by the facts of human biology.

Hence the biologist and sociologist can not but be interested in

[1] W. R. Inge, the former Dean of St. Paul's, did his best to foster this superstition of course, by publishing his own "family tree" (in *Outspoken Essays*, II. 260). But even in cases where achievements more valuable to the community than orthodox theological scholarship recur in close successive generations, the point is that this is the rare exception, not the rule.

the most widely voiced criticism of modern democracy, namely that it is inefficient as compared with totalitarian or authoritarian rule. This criticism is not one which can be waved aside. Those of us who have had experience of existing national organisations such as the labour movement, realise only too well the creaking slowness with which such bodies react to the stream of events. I knew a trade unionist once who only consented to become secretary of his local Trades Council on condition that he was not obliged to instal a telephone at his house, fearing too many calls on his time; and this in a period when Europe was being shaken to its foundations by the aggressions of the fascist powers. If an executive committee is accustomed to meet on the first Tuesday of every month, the heavens might fall before it would agree to be called together at an intermediate time. But all through evolution, organisms have been forced to resort to ever more efficient machinery for the control of their growing complexity. At the present time, human society has not yet developed the technique which it needs for looking after its affairs. The solution of the fascists, to abandon democratic principles, is a false one, since it involves treating human beings as units less highly organised than in fact they are. The proper solution is the "mechanisation" of democracy.

Perhaps I can explain what I mean by this phrase if I refer to the meeting of the International Physiological Congress in Moscow in 1935. At the plenary session, each delegate was supplied with headphones which he could plug in to one of a number of sockets in the desk before him. According to which one he chose he could hear the speech being delivered, either in the language of the speaker, or in French, German, English or Russian, spoken more or less sentence for sentence with the visible speaker, by hidden broadcasters from previously prepared translations. This is the kind of mechanisation of which we ought to have far more. There is nothing except prejudice and financial reasons which prevents each member of an executive committee, for example, being able to confer with the secretary at any time, possibly by small radio sets. These can be developed well enough for the purposes of war but so far have been not applied for making the machinery of democracy work in peace-time. In his book *The Social Relations of Science*, J. G. Crowther has written:

> "At present science is developing in the direction of big instruments and organisations. It seems probable that it will

evolve through this phase, and arrive at a new and higher one in which its instruments will again be small and compact. Science may show how a man can provide all his needs, in communication, food and transport, from very small instruments and concentrated supplies carried in his pockets.... If science is developed to this stage, it would provide new concrete bases for freedom."[1]

The Dominance of Economics over Eugenics.

In the case of an animal occupying a known oecological niche, there is but one possible environment, and the dominant species may be said to be well fitted for it. But the environment of man is not given *a priori*. In an exceedingly important paper, the great geneticist H. J. Muller,[2] has once and for all demonstrated the falseness of the assumption often made, that our present social stratification is correlated positively with genetic worth. Apologists for the existing social order, he says, defend their position with the *a priori* argument that in the social struggle, the better rise to the top. They neglect to show that success in modern economic competition depends on many factors beside innate endowment, and that to-day we have increasingly operative the principle of "to him that hath, more shall be given." But let us assume that inborn differences do play some role. The question is, what role? Are the characteristics which now lead men to rise economically, those which are the most desirable from a social point of view? It could at least as well be maintained that the dominant groups tend to have the genetic equipment which would be least desirable in a well-ordered social system, since they have been selected chiefly on the basis of predatory, rather than constructive and unselfish behaviour. This opinion is confirmed by a view of the lives of the more eminent financiers. It may even be maintained with some show of reason that the successful captain of industry, the eminent military leader or politician, and the successful gangster, are psychologically not far apart. In the social battle of present-day society, the high-minded, the scrupulous, the idealistic,

[1] p. xxvii.
[2] H. J. Muller, Sci. Monthly, 1933, **37**, 40. Muller is one of the most brilliant pupils of T. H. Morgan. It was Muller who discovered the effect of X-rays in increasing the mutation-rate of animals—a discovery which, apart from being an important research tool, may still open the way to an understanding of the mechanism of evolution itself. He was for long the director of the biological departments at the University of Texas, and later he organised the famous genetics institute at the Academy of Sciences in Moscow.

the generous and those who are too intelligent to wish to confine their interests to their personal monetary aggrandisement, are apt to be overwhelmed. This is what Muller calls the "Dominance of Economics over Eugenics." But what scientific evidence there is available, tends on the other hand to show that the differences in scores on intelligence tests, made by different races and classes, are, to the best of our knowledge, due to the differences in environmental advantages which they have received. It must, of course, be admitted that as yet we are ignorant of the distribution of genetic worth in the population of any civilised country with a capitalist form of economics and a strong class-stratification. Hogben has shown in an elegant way[1] that while we may safely speak of a genetic difference between two groups measured in one and the same environment, or between two environments in which one and the same genetic group is tested, we cannot discuss the relative contributions of inheritance and environment in the case of different groups in different environments. Those of us who propose to abolish class-distinctions and privileges must therefore act in faith, a faith supported, it is true, by the results of the tests mentioned above, the faith that the contribution of genetic worth at present suppressed by adverse environment, will be immensely large and significant. We shall not be deterred by this fact, for here humanity has nothing to lose.

Muller returned to the attack in a small but important book[2] in which he ranged over the whole question of the past, present and future of man, arising "out of the dark" of prehistoric savagery into more or less civilised communities and possessing a potential future of great grandeur. Man must recognise, he said, that this future can only be brought about by the continued and consciously fostered growth of that intelligence and co-operation which have brought him to his present status. Similar views have been forcefully expressed by another geneticist, Mark Graubard.[3]

Both these biologists agree that the social struggle under our existing civilisation does not lead to the reproductive survival of the germ plasm most favourable to this intelligence and co-operation. A competitive civilisation based on predatory individualism must, and obviously does, give particularly good opportunities to those persons

[1] L. Hogben, Journ. Genetics, 1933, 27, 379.
[2] *Out of the Night* by H. J. Muller (Vanguard Press, New York, 1935; Gollancz, London, 1936).
[3] *Genetics and the Social Order* by Mark Graubard (Tomorrow Press, New York, 1935) and *Man the Slave and Master* (New York, 1938 and London, 1939).

whose genetic constitution predisposes them for aggressive action.[1] Yet, as we have seen, humanity must move onwards to closer co-operative forms or perish. Individuals of the opposite type, characterised by more than the usual tendency to affection, kindliness, and fellow-feeling, naturally adapted to a co-operative social order, are on the contrary *selected against*. In the world of commerce, they have to go out of business; in learned circles, their just recognition is stolen from them; in medicine they are worn out with over-work and under-pay. Most significant of all, into the higher ranks of administration, government, and finance, they never penetrate. This state of affairs requires a "godly thorough reformation." After the adoption of an overtly social ideal by society, says Muller, and the changes in economics, ethics, and education that this will bring about, the urge to utilise all means of progress towards this ideal will become irresistible. The possibility of the genetic as well as the environmental improvement of mankind will then for the first time receive adequate recognition. But this will not in the least resemble the programmes of those present-day eugenists who accept the class-stratification of our society as inevitable and naïvely equate it with assumed genetic differences.

The Denial of the Sociological Level; "Physicism."

If some thinkers try to force the phenomena of human social life into the narrower framework of purely biological categories, it is perhaps hardly surprising that others should have made the attempt to force them into categories of physico-chemical type. This is what Pareto did. The *Treatise on General Sociology* of Vilfredo Pareto, first published in 1916, has attracted attention in much wider circles than those of professional sociologists and economists. It has been described as a work of genius which should be read by all who take an interest in human affairs and human relationships, whether politically or from the point of view of the pure spectator.

Pareto was born in 1848 in Paris, where his father was in exile on account of revolutionary activities in connection with the party of Mazzini. It was against this revolutionary atmosphere that Pareto was in rebellion throughout his life. He shared in the disillusionment

[1] The Tory propagandist, F. J. C. Hearnshaw, has the following extraordinary passage in a pamphlet issued by the Individualist Bookshop, Ltd. (No. 7). "Socialism is the system which legislates unsuccessful people into prosperity by legislating successful people out of it. It is essentially predatory." He has not examined his criteria of success; or perhaps they hardly admit of statement in plausible or even diplomatic terms.

which followed the victory of the Italian national cause in 1866, but instead of being temporarily alienated from the ideals of his father's generation, he retained to the end a positive hatred for them. He is continually attacking humanitarianism, the "god of progress," etc. Trained as an engineer, he occupied a high post in the Italian railways, and was for a long time active in the effort to induce the government of the day to adopt the principle of free trade, which he believed to be an essential condition of economic prosperity. This led to a conflict with the government so severe that he had to go into retirement. Before long, however, he was invited to occupy the chair of economics at the University of Lausanne, and from 1894 to 1923 he lived at Celigny "shutting out the troubles of the world, cultivating and storing the finest wines and fruits, and enjoying the material and spiritual pleasures of life." It is striking that one who has been called the chief theoretician of Italian Fascism should have been an aristocrat by birth, the son of a revolutionary father by upbringing, a disappointed politician in middle life, and a sybaritic Professor in old age.

How did Pareto go to work? L. J. Henderson (himself one of the greatest living biologists) shows[1] that he tried to apply to the social sciences, where the variable factors are the natures and interactions of human beings, the concepts of equilibrium which have been found so essential in the physical and biological sciences. First of all, however, Pareto's treatise is not "normative"; that is to say, he is concerned with "what men do, and not with what they ought to do." He is interested in the concept of the social system. His social system contains individuals roughly analogous with the "components" of the thermodynamic physico-chemical systems of Willard Gibbs. It is heterogeneous, i.e. in physico-chemical terms, it contains various "phases," for the individuals are of different families, trades, and professions, associated with different institutions and members of different economic and social classes. And as Gibbs considers temperature, pressure, and concentrations, so Pareto considers "sentiments" or, strictly speaking, the manifestations of sentiments in words and deeds; "verbal elaborations," and economic interests. The analogy with the Gibbs system drawn by Henderson is illuminating, but neither he nor other commentators, such as Borkenau,[2] seem to have sufficient acquaintance with the work of Marx and Engels. If they

[1] *Pareto's General Sociology: A Physiologist's Interpretation* by Lawrence J. Henderson (Harvard University Press, Cambridge, Mass., 1935).
[2] *Pareto* by Franz Borkenau (Chapman & Hall, London, 1936).

had studied this as carefully as that of Pareto, the suggestion could hardly have been made that Pareto was the first to elaborate the concept of the social system as an equilibrium mixture, alteration of which at one point duly affects all the other points.

But where the system of Pareto differs most profoundly from that of Marx seems to be that Pareto's is an "idealistic" as opposed to a materialist one. Pareto pays a great amount of attention to the first two factors mentioned above; the sentiments and the verbal elaborations, which he calls respectively Residues and Derivations. He is therefore much more of a psychologist than Marx, for whom all ideological "superstructure" is secondary, though in its turn reacting back. Translated into psychological terms, Residues would be complexes, and Derivations would be rationalisations, but the translation would leave a good deal to be desired. Residues are well explained by Henderson as follows: for ceremonial purification some peoples have used blood, the ancients used water, and water is still used in christian baptism to symbolise the effacement of sin. In these phenomena there are manifested at least two sentiments, the sentiment of integrity of the individual, and the sentiment that actions favourable to this integrity can be performed. These are Residues. They may be thought of as the residuum left after all the variable features of the phenomena have been dissected away. But the phenomena also include explanations of these ritual processes. These are Derivations. An extremely large part of Pareto's book is devoted to analysing and classifying residues and derivations. He evidently regards them not as secondary effects of changes in the productive economic relationships of men, but as primary causative agents in human mass actions.

The interest of this classification and analysis can hardly be disputed. But most of its details are disputable, as Borkenau demonstrates, and it is when we pass to other aspects of Pareto's work, such as his theory of élites and his treatment of utility, that the extreme disadvantages of his purely idealist approach make themselves felt. His theory of élites is completely vitiated by the improbable assumption that the class-stratification of modern European capitalism corresponds to some biological or genetic differences in the classes concerned. Borkenau's chapters on this subject are especially damning. Pareto assumes, moreover, as a basic point of his analysis, that class-domination *must* exist, since the special demands of a given society will lead to special treatment for those who possess the special abilities it most requires. This could not for a moment be accepted by those who advocate the communal ownership of the means of production. Class-domination is meaningless in the absence of property, through which alone hereditary classes can be perpetuated. Then secondly, his treatment of utility, though it involves scrupulous classification, is yet somewhat naïve. The distinction is made between the utility *of* a collectivity (people, nation, state) and utility *for* a collectivity. Consider the population of a country. There will be a certain optimum population. But optimum for what? A great increase of population will increase the "utility" *of* the country by leading to increased military and political power. For this the optimum would be high. A lesser increase in population, however, might lead to a maximum distribution of individual goods (utility *for* the country). Hence this optimum would be much lower. But is this more than a rather pedantic way of saying that an imperialistic foreign policy demands a large internal population as potential cannon-fodder? This we knew before. Everyone has noted the double activity of Mussolini and Hitler in discouraging birth-control and demanding territorial expansion at one and the same time.

The upshot of the matter is that although the biological, and even the physico-chemical, approach to sociology has a certain part to play, any attempt to construct a sociology on their foundations alone is bound to lead to failure. Pareto was not the first to have recourse to the analogies of physico-chemical equilibria; Bogdanov and others in Russia had taken that line at the beginning of the century, and it is interesting that they were severely criticised by Lenin.[1]

The Contribution of Psychology.

So far this review has mainly been concerned with the contribution of Genetics to a human biology. But there are reasons for thinking that Psychology, Physiology and Biochemistry have equally great contributions to make. It seems that sociologists have hitherto somewhat underrated the possible value of a psychological approach, repelled perhaps by the exaggerations of some of the lesser followers of the psycho-analytic schools. Nevertheless, I am convinced that on most of the fundamental problems, the psycho-analysts are correct in their attitude. The psychologists of the future ought to be able to provide us with a theory of child upbringing which would abolish, by a kind of preventive medicine, the warped and unhealthy mental states which to-day so often threaten the otherwise good relationships

[1] In *Materialism and Empirio-Criticism*, Works, Vol. II, p. 379.

which might be established between human beings. Revolutionaries, it is true, are perfectly right when they complain that many psychoanalysts aim at reconciling their patients to the existing order, i.e. of fitting them for their environment; and that they do not consider the possibility that this environment is itself sick.[1] But the proper conclusion surely is that both the social environment *and* man himself need radical alteration.[2] To abandon the attempt to transform the former would be to go back on all the past centuries of human struggle. Yet whatever we do to it, there is no immediate possibility that we could make man any other than an organism beginning by growth and development *in utero*, having parents and later developing mentally under their influence or the influence of substitutes for them, and at last living in a space-time continuum with all the inevitable frustrations depending upon this fact. There is, as it were, an irreducible minimum of hardness in man's condition. Take, for example, Eliot's poem, "Animula:"[3]

> "The heavy burden of the growing soul
> Perplexes and offends more, day by day;
> Week by week, offends and perplexes more,
> With the imperatives of is and seems
> And may and may not, desire and control,
> The pain of living and the drug of dreams
> Curl up the small soul in the window-seat
> Behind the *Encyclopaedia Britannica*."

Or the speech of the Four Tempters to Thomas of Canterbury in the play:[4]

> "Man's life is a cheat and a disappointment;
> All things are unreal,
> Unreal or disappointing;
> The Catherine wheel, the pantomime cat,
> The prizes given at the children's party,
> The prize awarded for the English Essay,
> The scholar's degree, the statesman's decoration...."

[1] Some psychologists do apparently realise this, e.g. Burrow and Frank; see Syz, Amer. Journ. Sociol., 1937, 42, 895.
[2] Two books from this viewpoint which are much to be recommended are *Soviet Russia Fights Neurosis* by F. E. Williams (London, 1934) and *The New Road to Progress* by S. D. Schmalhausen (London, 1935).
[3] T. S. Eliot, "Animula" in *Collected Poems*, p. 111.
[4] T. S. Eliot, *Murder in the Cathedral*, p. 44.

For stability and happiness, then, in the most just social order which we can conceive, psychological hygiene will at all points be necessary. With the "liquidation" (as we might say) of the neuroses, paranoia, and the generally unrecognised but potent minor conditions, such, for example, as sadism, which existed beforehand, an enormous part of human unhappiness will disappear. I am well aware of the fact that psychologists are often in diametrical disagreement as to the proper means to be adopted for this end; e.g. sexual frustration is held by Kyrle[1] to be a force adverse to civilisation, and by Unwin[2] to be precisely the opposite. But in general I believe that the movement initiated by Freud has promise of sociological importance wider than most of us can yet realise, and that criticisms of such books as that of Glover[3] on the psychologic causes of war, from the point of view of the primacy of economic causes, are misdirected though not unjustified.

The problems of sex illustrate well how the reformation of individual life follows upon the reformation of social life. As Schmalhausen[4] says:—

> "Property, [modern] puritanism, and pornography go surprisingly well together. Property is concerned with institutions, not with impulse. Puritanism is concerned with tabus on impulses dangerous to the patriarchial-propertied way of life. Pornography is the safety-valve for emotions and curiosities repressed by patriarchal-propertied puritanism. Under these darkening and constricting auspices, sex was a reality remote from love, and love dissociated from marriage. . . . Since the love of comrades, rooted in sex attraction, but flowering in sheer humanness, will be the natural state of affairs in our new society, intimacy will know a wider range of possibility and fulfilment than was admissible under repressive civilisations, except in secret forms of 'immorality' and perversions. A certain aboriginal innocence, out of primitive and pagan patterns, will be restored to sex love, giving it wings and inspiration once more among the pleasures and satisfactions of life. Christian marriage has been built on three weird assumptions; that sex is evil, that love is unreliable, and that unwanted children are socially desirable. Such marriage is the compulsory method of coercing man's

---

[1] R. M. Kyrle, Psyche, 1931, **11**, 48.
[2] J. D. Unwin, *Sex and Culture* (Oxford, 1934).
[3] E. C. Glover, *War, Sadism, and Pacifism* (London, 1933). [4] Loc. cit., pp. 237 ff.

evil impulses into giving women unwanted children, for the greater glory of God. Communist marriage drastically alters these assumptions; it innocently assumes that sex is good, love attainable, and marriage with or without children the comradely creation of a man and woman who enjoy each other's company so much that they want to try the experiment—it can not rationally be more than that—of living together for that period of time during which their intimate friendship makes life seem more livable and lovable. In communist ethics, the unholy alliance of property, puritanism and pornography is dissolved, sex love becoming candid without lewdness, informal without degradation, free without debauchery."

In this connection one cannot but be reminded of those sparks of prophetic insight concerning human relationships which scintillate among the apocryphal writings of early christianity.[1] In the *Gospel according to the Egyptians*, which Origen criticised, there occurs the following:—"When Salome inquired as to when the things concerning which she asked should be known, the Lord said: 'When you have trampled on the garment of shame, and when the two become one, and the male with the female is neither male nor female." A similar passage occurs in the *Second Epistle of Clement*.

In the fragment of a gospel in the Oxyrhynchus Papyrus (655) there occurs the following: "His disciples said unto him, 'When wilt thou be manifest unto us, and when shall we see thee?' He saith, 'When ye have put off your raiment and are not ashamed.' "

In reading these strange fragments one cannot help thinking of the great advance in civilisation which is implied by the possibility of men and women working together in intellectual and productive work. For earlier forms of civilisation, in which sex was a force far less under rational and technical controls than now, this collaboration was impossible. The liquidation of male aggressiveness and the liberation of the female from age-old slavery is the guarantee of the stable and organised society of the future with its infinite control over nature.

The Contribution of Physiology and Biochemistry.

The contribution of physiology to human affairs may seem to many to be exhausted by its effects upon our knowledge of what is healthy and what is not. Thus the realisation that Vitamin D is

[1] The quotations following are from *The Apocryphal New Testament* (Oxford, 1924).

THE BIOLOGICAL BASIS OF SOCIOLOGY

synthesised in our skin by the photo-chemical action of ultra-violet light leads to discontent with the conditions of life in our cities and hence to means of combating them, such as more sensible clothing, or lack of clothing, and provision for quick transit to and from urban and industrial areas. But nerve-physiology has far more fundamental fruits than these. In a particularly striking and irrefutable manner, such researches as those of Pavlov[1] and other neurologists show the error of the ancient aphorism that "human nature cannot be changed." Sociologists should have no further connection with theories which postulate any such unchangeableness. The whole group of facts centering round the phenomenon of the conditioned reflex means that the nervous system of the higher mammal is still pliable; that new reflex paths may be established; that because we are, to a large extent, what our education (in the widest sense) was, by remodelling that education, the men and women of the future may be as superior to us as we are to the painted Britons. We cannot too much insist on this contribution of neurology to sociology. In so far as sociology is normative, it must seek means to increase the general good. For this purpose it has always been customary, largely because of the influence of individualistic christian theology, to accept the view that amelioration must begin in the individual unit, and thence spread outwards to the mass of humanity. But nerve-physiology seems to show that just the opposite is true. Radical improvement in the social conditions governing education would condition the mass of humanity to the higher levels of comradeliness and social cohesion necessary for the more highly organised, and hence ameliorated, state of society. It must, however, be admitted that this outcome of physiology works both ways. "Conditioning" may be used for good purposes but also for bad ones. We need only refer to the millions of young people now being conditioned in fascist countries to the notion that racial struggle is inevitable in the world, that the racial or national flag is the highest symbol of human effort, and that war is the most natural activity of man. This conclusion of physiology receives further support from anthropology. That human behaviour ("human nature") may be inconceivably different according to the social structure of the unit in which it is displayed, appears with great force in such a book as that of Mead,[2] where the sex behaviour of three different primitive peoples is contrasted in detail.

[1] I. P. Pavlov, *Conditioned Reflexes* (Oxford, 1927, tr. G. V. Anrep).
[2] M. Mead, *Sex and Temperament in Three Primitive Societies* (London, 1935).

Lastly, as in private duty bound, I come to the subject of biochemistry. To sociologists the subject of human nutrition cannot fail to be of interest, and it is biochemistry which shows us what a properly-balanced and adequate diet should contain, and how any given diet falls short of these requirements. We shall doubtless all be familiar with the recent work of Orr[1] in which the nutritional aspects of the class-stratification of the social system in which we live, are graphically plotted. An optimum diet is defined as a diet to which nothing need be added to ensure a better degree of health and well-being. In England we find that about half of the entire population fall below this level, and clearly this must mean a great deal more than half of the working-class. In the lowest group of all, containing 10 per cent of the population, the diet is deficient in every single constituent, minerals, vitamins, organic substances, etc., and it is significant that no less than 25 per cent of the children of the country fall into this group, precisely because the larger number of children in a family, the more tendency there is for the family to come into the lowest group, even if the wage-earner is well paid relatively, since the groups are reckoned on a *per capita* basis of weekly income. In the disputes of recent times about the absolute minimum upon which a family can support life, we have seen biochemistry, brought into the limelight by august public bodies whose good intentions and would-be generosity cannot be questioned, become a subject as controversial as geology once was when the episcopate was a more powerful (or less diplomatic) intellectual force than it is now. Whatever the outcome may be, the sociologist must always take into account, as one of his essential base-lines, the knowledge concerning the chemical constitution of man and its maintenance, which it is the business of the biochemist to provide.

Boundaries of Social Theory.

In conclusion, to restate the thesis of the present paper would be to say that biology, like biochemistry and biophysics, is an unescapable datum for the sociologist. It gives the limits within which the answer must fall, and not the answer itself. Like the crystal physicist who can say of the proposed chemical formula of a substance that whatever the true one is it cannot be that one, without being able to say what the true one is; so the biologist can lay down numerous boundaries which no social theory can expect to transgress. Whatever the details

[1] J. B. Orr, *Food, Health and Income* (London, 1936).

of the just social order, it is not true that human nature cannot change, as our social inheritance develops; it is not true that human beings do not need more than fifty calories a day, or so many milligrams of such and such a vitamin; it is not true that psycho-analysis throws no light on human behaviour. The sciences of the sub-social organisational level state, in fact, the laws applicable to human organisms as animal organisms. But by virtue both of their humanity, and of the fact that they congregate together in supra-human organismic groups, they constitute a higher level of experience, for which the laws of the lower level alone will not suffice. The sociologist must cope with the greater complexity on its own ground. To treat sociological problems with purely biological concepts is to add oneself to the long list of humanity's false prophets.

A Biologist's View of Whitehead's Philosophy

(From the Whitehead volume in the Library of Living Philosophers, 1941)

THE author of this contribution is very conscious, both of the honour which has been done him in an invitation to contribute to the present symposium, and of the pleasure which it is to be able to participate in a reasoned tribute to one of the greatest of living philosophers. The reader's indulgence is begged at the outset in case any of what follows should appear unduly trite, but in requesting the co-operation of a working biologist, the editors laid themselves open to receiving a paper in which the finer points, to say the least of it, of Professor Whitehead's philosophy, should be but poorly appreciated.[1]

Trends in Theoretical Biology.

The author's interest in "philosophical" or theoretical biology was probably awakened by the very fact of his being a biochemist. The zoological systematist may get along well enough by treating his data as an array of empty forms unconnected with a material substratum; the psychologist may do the same; and the organic chemist may reveal the structural formula of some compound once involved in a living cell, or analyse the constituents of blood or tissue fluids, without devoting much thought to the organisation of the living being which synthesised the one or secreted the other. But the true biochemist is deeply concerned about the structure and organisation of the living cell, with its "topography" permitting of innumerable simultaneously-proceeding chemical reactions, its faculty of getting things done just at the right time and place, and its remarkable properties of symmetry and polarity, exhibited in an aqueous colloidal medium of certain essential constituents, especially the proteins,

---

[1] Abbreviations adopted in what follows:

   S & MW *Science and the Modern World.*
   AOI *Adventures of Ideas.*
   N & L *Nature and Life.*
   P & R *Process and Reality.*
   MOT *Modes of Thought.*

carbohydrates and fats. The ancient problem of body and mind, too, was always around the corner. How to reconcile the introspected "me" and the domain of mind and spirit with the world of flesh and blood, of macromolecules and hydrogen ions, with which the former seemed to be so strangely connected? This interest in the organisation of the living cell, the borderline between biology and physics, was natural in Cambridge, where the tradition of W. B. Hardy and F. G. Hopkins was, and still is, in full vigour.

The author's first approach to the whole subject on the theoretical side was therefore a careful, if somewhat unrewarding, examination of the voluminous and polemical literature on "vitalism," "neo-vitalism," and "mechanism," which had appeared during the last decade of the past century and the first two of the present one. The writings of Hans Driesch, J. S. Haldane and E. S. Russell on the one side, and of men such as Jacques Loeb, H. S. Jennings and Judson Herrick on the other, were gone through. It would be probably well worth someone's while to take this literature and make a coherent historical summary of it, for it belongs to a distinct period which has been closed since about 1930. The vitalists systematically drew attention to the flaws in the over-simplified explanations of biological processes which workers such as Loeb, recognising them to be interim hypotheses only, were always putting forward. Their attitude was no doubt partly inspired by the human, all too human, but nevertheless obscurantist, desire to retain elements of mystery in the universe, and hence they fought decade after decade a stubborn withdrawing action against the ever-fresh shock-troops of the mechanists. The process had begun long before; it was familiar to the men of T. H. Huxley's time, as witness the interesting passage in the book of that curious character, W. H. Mallock, *The New Republic* (1878).

> "*Saunders* (intended to be W. K. Clifford, the mathematician):
> 'One word more, one plain word, if you will allow me. All this talk about religion, poetry, morality, implies this—or it implies nothing—the recognition of some elements of inscrutable mystery in our lives and conduct; and to every mystery, to all mystery, science is the sworn, the deadly, foe. What she is daily branding into man's consciousness is that nothing is inscrutable that can practically concern him. Use, pleasure, self-preservation—on these everything depends; on

these rocks of ages are all rules of conduct founded, and now that we have dug down to these foundations, what an entirely changed fabric of life shall we build upon them. Right and wrong, I say again, are entirely misleading terms, and the superstition that sees an unfathomable gulf yawning between them is the greater bar to all health and progress.'

"*Laurence* (intended apparently to be Mallock himself):

'And I say, on the contrary, that it is on the recognition of this mysterious and unfathomable gulf that all the higher pleasures of life depend....'"

The vitalists, in fact, were concerned, perhaps because of some misplaced sense of the numinous, to retain at all costs a measure of animistic mystery in the nature and behaviour of living organisms. The neo-vitalists, as they called themselves, centred this mystery in the very organisation of life itself, regarding organising relations as in principle inscrutable and axiomatic, rather than a subject for investigation.

The atmosphere surrounding these controversies was always somewhat polemical. As W. T. Marvin said in 1918, compare Driesch and Loeb—nobody could call them unimpassioned neutrals examining a body of evidence. He went on to suggest a psychological difference; the vitalist hoped that the scientific method as applied to life and mind would fail, the mechanist hoped it would succeed.

"If science wins, the world will prove to be one in which man is thrown entirely on his own resources, skill, and self-control, his courage and his strength, perhaps on his ability to be happy in adjusting himself to pitiless fact. If science fails, there is room for childlike hopes that unseen powers may come to the aid of human weakness. If science wins, the world is the necessary consequence of logically related facts, and man's enterprise the playing of a game of chess against an opponent who never errs and never overlooks our errors. If science fails, the world resembles fairyland, and man's enterprise no longer a test of skill and knowledge, but conditioned by the goodness of his will or the possibility of luck."[1]

Vitalists and neo-vitalists were found rather among the philosophers than among the biologists themselves. Certainly during the present

[1] W. T. Marvin, Philos. Rev., 1918, **27**, 616.

century the vast majority of working biologists and biochemists have been "mechanists." Their conception of the task of biology was consistently that sketched out in T. H. Huxley's definition of physiology in 1867: "Zoological physiology is the doctrine of the functions or actions of animals. It regards animal bodies as machines impelled by certain forces and performing an amount of work which can be expressed in terms of the ordinary forces of nature. The final object of physiology is to deduce the facts of morphology on the one hand and those of oecology on the other, from the laws of the molecular forces of matter."[1] This, however, though useful as a slogan, can never have satisfied even the working biologists. It must always have been obvious that the laws of chemistry do not appear until you are dealing with entities sufficiently large to show the phenomena of chemical combination, and similarly that the laws governing crystal structure do not appear until crystals have been formed, and *a fortiori* the laws of living organisms or social units cannot be studied except at their own level. This is the problem which the emergent evolutionists afterwards brought into prominence. If one were to know *all* there is to know about the properties of atoms, for instance, it may be said, one should be able to predict all the molecular combinations they would form, and all the living structures that could be built up from them; but in order to know all about the atoms one has to know a great deal about the molecules and the living cells first. In 1838 K. F. Burdach had said "Physiology will always be able to dispense with the aid of chemistry." This was not necessarily a vitalist statement. During the succeeding century it became quite clear that the regularities established in the biological sciences—physiology, experimental morphology and embryology, genetics, cytology, and the like—remained of durable validity, whatever discoveries might be made in biochemistry and the organic chemistry of substances of biological origin, to say nothing of biophysics. The question became critical; how are the levels related? How do biochemistry and biophysics contribute (as they obviously do) to a unified picture of life and nature? For certain studies the problem was a desperate one. When Wilhelm Roux in the last two decades of the nineteenth century founded the science of experimental morphology ("Entwicklungsmechanik") by the strict application of causal analysis to developmental processes instead of their mere

[1] T. H. Huxley, Science Gossip, 1867, p. 74, and in *Lay Sermons; Addresses and Reviews* (London, 1887), p. 83.

description, he divided the biological factors into two, the "simple components," whose connection with physico-chemical factors could immediately be seen, and the "complex components," where the relation with physico-chemical factors was much less obvious, but might reasonably be expected to be revealed in due course.[1] The curling of a piece of ectoderm, for instance, if understood in relation to protein fibres, surface forces, lipo-protein monolayers, etc. would be a case of a simple component. The regularly reproducible self-differentiation of an isolated eye-cup under certain conditions, for instance, involving processes much too complicated at present for physico-chemical explanations, would be a case of a complex component. But the terminology of *components* (*einfache* and *complexe Komponenten*) never came into general use.

By 1928 the position of most working biologists could be summed up not unfairly as follows:

> "Mechanists do not say that nothing is true or intelligible unless expressed in physico-chemical terms, they do not say that nothing takes place differently in living matter from what takes place in the dead, they do not say that our present physics and chemistry are fully competent to explain the behaviour of living systems. What they do say is that the processes of living matter are subject to the same laws that govern the processes in dead matter, but that the laws operate in a more complicated medium; thus living things differ from dead things in degree and not in kind; they are, as it were, *extrapolations* from the inorganic."[2]

The nature of this relationship, however, still remained obscure. In the following year, however, the situation was greatly clarified by the appearance of J. H. Woodger's book *Biological Principles*. This important work set out to discuss the classical contradictions in biological theory, vitalism and mechanism, structure and function, organism and environment, preformation and epigenesis, teleology and causation, mind and body. Contradictions once recognised, fruitful synthesis followed. From his discussion it followed that the term "vitalism" ought henceforward to be restricted to all propositions of the type "the living being consists of an X in addition to carbon,

---

[1] See W. Roux, *Gesammelte Abhandlungen ü. Entwicklungsmechanik der Organismen* (2 vols. Leipzig, 1895).
[2] SB, p. 247.

hydrogen, oxygen, nitrogen, etc., plus organising relations." This, for biologists, at any rate, was one of the first clear statements of the objectivity and importance of organising relations in the living system. They had always been recognised, but at the same time obscured, by the persistent opposition of progressive experimental science, implacable but correct, towards every form of lingering animism, the *spiritus rector*, the *nisus formativus*, the *archaeus*, the *entelechia*, etc., etc. And the situation had not been improved by the adoption of the organising relations by the neo-vitalists as the very citadel of the *anima* itself.

Organising relations, then, were to become the object of scientific study, not the home of an inscrutable vital principle, nor the axiom from which all biology must proceed. Since 1930 this point of view has penetrated widely through scientific circles, all the more so as it was really a description of what a large number of scientific workers had previously believed in a somewhat unconscious way. If space permitted, it would be interesting to consider the practical applications of these ideas, the question, for instance, of what methods may be adopted in the study of organised living structure. How far may wholes be made transparent, as by X-ray analysis of the crystalline and liquid crystalline arrangements which, as we now know, play such an important part in the structure of the living body? How far can living structure be so explored without interference with its delicate organisation (cf. the principle of indeterminacy)? What are the forces which hold morphological entities together, and how do they link up with forces at the molecular and sub-molecular level? We cannot go into these questions at this time. Biological organisation is not immune from scientific enquiry, it is not inscrutable, and it cannot be "reduced" to physico-chemical organisation, because nothing can ever be reduced to anything. As Samuel Butler once remarked, "Nothing is ever *merely* anything." The laws of higher organisation only operate there.

For the emergent evolutionists, as is well known, emergence was a logical as well as a historical category. Not only had the various levels of complexity in the universe emerged in a historic sequence; but each was logically unpredictable from the basis of those lower than itself. Modern physical accounts of chemical events and chemical accounts of biological events, however, have rendered this point of view too simple. Chemical behaviour can be deduced from atomic physics, and biological behaviour from biochemistry. But the essential point is that by the time this has become possible, the physics and the chemistry have been completely transformed by the incorporation of wider factual ranges. The matter has recently been well put as follows: "It seems to become clear that chemistry will never become predictable mathematically, and that, in fact, we have rather to make mathematical physics—in a sense—more chemical. We have to discover a set of empirical simplifications—corresponding to the nature of the chemical properties of matter—which will allow us to crystallise the general equations of atomic physics into a form readily applicable to chemical changes."[1]

Woodger's point of view, supported also by the Austrian theoretical biologist v. Bertalannfy,[2] was perhaps the most technically well-informed manifestation of a great movement of modern thought which sought to base a philosophical world-view on ideas originating from biology rather than from the classical physics. It fused once again what Descartes had put asunder. It was Descartes, as Woodger acutely said, who introduced the practice of calling organisms machines, with the unfortunate consequence that transcendent mechanics had to be invented to drive them. Organicism, if not obscurantist, was bound to be the death of "vitalism" as well as of "mechanism." It was likely to be the death of animism too, since mental phenomena cannot but be considered in the light of the evolution of the central nervous system, as is discussed in the latest and most interesting work of this kind, the Gifford Lectures of Sir Charles Sherrington.[3]

Succession in Time and Envelopes in Space.

From the scientist's standpoint, the organic conception of the world involves *succession* in time and *envelopes* in space. Taking the latter first, it is obvious that the different levels of organisation, for such we must call them, occur one within the other. Ultimate particles, the proton, electron, etc. build up atoms, atoms build molecules, molecules build large colloidal particles and cell-constituents and paracrystalline phases and the like, these in their turn are organised into the living cell. Above this level, cells form organs and tissues, the latter combine into the functioning living body, and the bodies of animals, especially men, form social communities. As the central

---
[1] M. Polanyi, Nature, 1942, **149**, 510.
[2] L. v. Bertalannfy, *Theoretische Biologie* (Berlin, 1932).
[3] C. Sherrington, *Man on His Nature* (Cambridge, 1941).

nervous system becomes more complex so mental phenomena emerge, until the elaborate psychological life of man is attained. There is a sense in which minds include and envelop bodies, for the boundaries of thought are far wider than those of what the special senses can record, and minds interpenetrate as bodies cannot[1] (see on, where Whitehead's concept of the focal region is described). The remarkable thing about our world is, however, that these envelopes seem each to be analogous to past phases in the history of its development. There were "inorganic" molecules before there were living cells, the origin of which evidently depended upon the right environmental conditions for the flowering of the potentialities of the protein system, there were living cells before there were organs or tissues of metazoan organisms, there were primitive organisms before there were any higher ones, and higher organisms before there were any social associations. The fundamental thread that seems to run through the history of our world is a *continuous rise in level of organisation*. Whether this organisation is the same as that to which physicists refer in their discussion of the shuffling process which underlies the second law of thermodynamics, and whether its rise in the domain of living organisms has entailed some corresponding loss of organisation somewhere else, are matters which we cannot stop to deal with here.

The basically important fact that social evolution must be regarded as continuous with biological evolution was appreciated already by Herbert Spencer, who in this respect, though not of course in others, made an approach to the organic conception of the world. It has the extremely important corollary that any static or too conservative view of the present position of human institutions becomes impossible. If living organisation has such triumphs behind it as the first invention of the cell-membrane, the kidney-tubule, the notochord, the flint-knife and the plough, the art of language and the skill of ships, it is not likely that the agreements of Ottawa or Munich have any durable importance, or that human society will always remain separated into states with national sovereignties above the moral law, and social classes with different privileges and manners. This has generally been appreciated by upholders of the organic view of the world, but much more boldly by Marx and Engels, for instance, than by Smuts, Lloyd-Morgan, or Sellars.[2]

[1] Though some of them sometimes long to.
[2] Smuts, J. C., *Holism and Evolution* (London, 1926); Lloyd-Morgan, C. (Gifford Lectures, 1st Series, *Emergent Evolution*, London, 1923, 2nd series, *Life, Mind and Spirit*, London, 1926); Sellars, R. W., *Evolutionary Naturalism* (Chicago, 1922).

It is probable, indeed, that the organic view of the world has considerable historical and social significance. The seventeenth century, the age of Gassendi and Newton, of Boyle and Descartes, was a time in which the capitalist system of economic individualism won its first decisive victories in taking over state power. The surrender of the last royalist troops in the English civil war was the final conclusion of centuries of feudalism, for though the monarchy was restored in England, feudalism was not. All later monarchs ruled by the grace of the City of London. Parallel with the breaking-up of the old guilds, and the absolute freeing of commercial enterprise in every kind of new exploitation, went the rediscovery of atomism by Gassendi and its application to chemistry as the "corpuscularian or mechanical hypothesis" by Boyle. The analogy between free merchants, projectors, and industrialists, and the fortuitous concourse of atoms, can even be found explicitly stated in seventeenth-century books on economics. Is it not therefore of interest that in our time, when capitalist economics has worked itself through to a new state of society demanding everywhere more social control and organisation of human affairs, that there should be a rediscovery of the organic interpretation of the world, an interpretation in which the monads "do not blindly run" in Whitehead's famous phrase,[1] "but run in accordance with the whole of which they form a part." Function depends on position in the whole. Statistical regularity of fortuitous random motions is not the whole story; there is a plan of organising relations too. The world is not entirely like a perfect gas or an absolutely homogeneous solid, it also contains viscous phases, crystals rigid in one, two or three dimensions, plasticity and elastic deformability, living organisation. It may be that we are on the threshold of a long period, lasting perhaps for several centuries, in which the organic conception of the world will transform society, giving it a unity more comradely and equal than feudalism, but less chaotic and self-contradictory than the centuries of capitalist atomism.[2] In Alfred North Whitehead we surely have to recognise the greatest living philosopher of the organic movement in philosophy and science.

Historical origins of Organicism and Dialectical Materialism.

About the historical origins of the organicistic viewpoint in biology a great deal could be said. There is space to refer only to one or two

[1] S & MW, p. 113.
[2] Or, as a friend of mind acutely puts it, Contract is not returning to Status but going forward to Function.

points. Samuel Taylor Coleridge is not generally regarded as having contributed much to theoretical biology, yet surely his essay *The Theory of Life*, published in 1848, was more advanced than any other thought at the time, and than a great deal since. Coleridge wrote:

> "I define life as the principle of individuation, or the power which unites a given all into a whole that is presupposed by its parts. The link that combines the two, and acts throughout both, will, of course, be defined by this tendency to individuation. Thus, from its utmost latency, in which life is one with the elementary powers of mechanism ... to its highest manifestation ... there is an ascending series of intermediate classes, and of analogous gradations in each class. ... In the lowest forms of the vegetable and animal world we perceive totality dawning into individuation, while in man, as the highest of the class, the individuality is not only perfected in its corporeal sense, but begins a new series beyond the appropriate limits of physiology."[1]

It is curious to think that Coleridge was as unconscious as Aristotle (who also recognised a "ladder of beings") of the evolutionary succession which has coloured all our thought on these subjects since the middle of the last century.

More important, some decades later, was the work of the London philosopher, Karl Marx,[2] and the Manchester business man, Frederick Engels.[3] The views of the latter on scientific theory have in recent times become generally recognised as having been far in advance of his age. The author would disclaim any competence for presenting the contributions of these great thinkers as they deserve, but there are numerous handbooks which may be consulted, a process which is in this case especially necessary as the views of these men on political subjects, then unorthodox, caused them to be somewhat boycotted in academic circles.[4] Marx and Engels were, of course, profoundly influenced by Hegel, just as Coleridge had been. But whereas he

---

[1] S. T. Coleridge, *Theory of Life* (1st edn., 1848, usual edn., London, 1885).

[2] See especially *The German Ideology* and *Theses on Feuerbach*.

[3] See especially *Anti-Dühring; Socialism, Utopian and Scientific*, and the *Dialectics of Nature*.

[4] See e.g. T. A. Jackson, *Dialectics, the logic of marxism*, 1936; D. Guest, *Textbook of dialectical materialism*, 1939; R. Maublanc, *La Philosophie du marxisme et l'enseignement officiel*, 1936; J. Lewis, *Introduction to Philosophy*, 1937.

tended to retain Hegel's metaphysical idealism, they "turned it right side up" and while keeping Hegel's dialectical account of change and process, in which a synthesis arises out of the deadlock of the thesis and the antithesis, they adopted a realist metaphysics. Their materialism, however, was to be known as "dialectical materialism" as opposed to the old mechanical materialism, in order to show its naturalistic character, its determination to account for all the highest phenomena of mind and social organisation without leaving the firm basis of the real objective existence of matter. Something of this kind was meant by Marx's saying that materialism must cease to be "misanthropic." Of course, the only way in which such a naturalism could account for the highest phenomena of mind and social organisation, love and comradeship, justice and mercy, was by admitting a series of levels of organisation, arranged in the successions and envelopes of which we have already spoken. And so from this standpoint also there came a doctrine of levels of organisation. It had, however, the cardinal virtue, which not many other naturalisms have had, of emphasising the transitory character of human institutions. It showed that evolution of social systems continued from that of biological systems, and urged the optimistic but tolerably convincing view that human misery is essentially connected with a low and inferior stage of social organisation, that it had in the past been much worse than it is now, and that in the future it ought to be greatly decreased. This is not the place to discuss Marx's theory of history, but if history is the history of class-struggles (and to some extent it undeniably is), there is room for hope that when mankind has united in a world co-operative commonwealth unmarked by social classes, a good many of the more unpleasant features of life in a semi-barbarous state will have ceased to exist. And indeed this is not a hope at all, but a prediction based on that guiding thread of rise in level of organisation, which we have seen running throughout the evolution of our world; and hence a scientific prediction. It was for this reason that the kind of socialism advocated by Marx and Engels received the name which it bears to this day, "scientific" socialism, as opposed to the Utopian varieties, which based their hopes only on the goodness of human nature or similar more or less reliable factors.

Dialectical materialism has been called the theory of transformations, of the way in which the qualitatively new arises, of the nature of change in the natural world. Its outcome in biology—to return to our main theme—was certainly beneficial. In 1931 a Russian biologist,

B. Zavadovsky, expounded this for an English symposium.[1] An equally good description was given by M. Shirokov:

> "A living organism is something that arose out of inorganic matter. In it there is no 'vital force.' If we subject it to a purely external analysis into its elements we shall find nothing except physico-chemical processes. But this by no means denotes that life amounts to a simple aggregate of these physico-chemical elements. The particular physico-chemical processes are connected in the organism by a *new form of movement*, and it is in this that the quality of the living thing lies. The new in a living organism, not being attributable to physics and chemistry, arises as the result of the new *synthesis*, of the new *connection* of physical and chemical movements. This synthetic process whereby out of the old we proceed to the emergence of the new was understood neither by the mechanists nor the vitalists. ... The task of each particular science is to study the unique forms of movement characteristic of that particular level of the development of matter."[2]

A few years later there was another good statement from a French biologist, professor of zoology at the Sorbonne. Marcel Prenant wrote:

> "In biology dialectical materialism is opposed both to vitalism and to mechanical materialism, which are both really metaphysical theories. He refuses to make a sharp distinction between the physical and biological sciences, to reserve causal determinism to the former and to appeal to teleology in the latter. But neither does he suppose that biology must try to reduce itself to the physical sciences. He affirms the unity of the world, in which neither life nor human society constitute domains apart, but he also affirms that this unity expresses itself in qualitatively different forms of whose distinctive characters one should never lose sight."[3]

Dialectical materialism has been perhaps more successful in emphasising the existence of the levels of organisation and in showing

---

[1] B. Zavadovsky, essay in *Science at the Cross-Roads* (London, 1931).
[2] From M. Shirokov & J. Lewis, *Textbook of Marxist Philosophy* (London, n.d.) p. 341.
[3] M. Prenant, Bull. Soc. Philomath. Paris, 1933, **116**, 84.

the dialectical character of human thought and discovery than in elucidating the dialectical character of the transitions between the natural levels.[1] There have, however, been some interesting suggestions. J. D. Bernal[2] has pointed out that natural processes are never 100 per cent efficient. Besides the main process or reaction, there are always residual processes or side-reactions, which, if cyclic or if adjuvant to the main reaction, will not matter very much. But they may be opposing and cumulative, so that after some time a new situation will arise in which such opposing processes may make an antithesis to the main reaction's thesis. This situation may be unstable, and wherever instability occurs, one of the possible resulting syntheses may be a level of higher organisation. Such a scheme can be worked out for the aggregation of particles in planets, the formation of hydrosphere and atmosphere, and the development of economic processes since the renaissance. J. B. S. Haldane,[3] too, has discussed evolution theory from this viewpoint, distinguishing three Hegelian triads:—

Thesis	Antithesis	Synthesis
(1) Heredity	Mutation	Variation
(2) Variation	Selection	Evolution
(3) Selection of the fittest individuals	Consequent loss of fitness in the species	Survival of those species showing little intraspecific competition.

The early conviction of Engels that Nature is through and through dialectical was rightly directed against the static conceptions of the scientists of his time, who were unprepared for the mass of contradictions that science was about to have to deal with, and who did not appreciate that Nature is full of apparently irreconcilable antagonisms and distinctions which are reconciled at higher organisational levels. The well-known rules of the passing of quantity into quality, the unity of opposites and the negation of negations, have all become commonplaces of scientific thought.[4] What has not yet been done, however, is to elucidate the way in which each of the new great

[1] Cf. F. Engels, *Dialectics of Nature* (Gesamtausgabe edn., Moscow, 1935), p. 640.
[2] J. D. Bernal, essay in *Aspects of Dialectical Materialism* (London, 1934).
[3] J. B. S. Haldane, Science and Society, 1937, **1**, 473.
[4] J. B. S. Haldane, *Marxism and the Sciences* (London, 1938), and a series of articles in Labour Monthly, 1941, **23**, 266, 327, 400 and 430.

levels of organisation has arisen, and although this must to some extent await the results of further experiments and observations, there is quite enough knowledge already available to permit of a good deal of theoretical thinking along these lines.[1]

One outstanding example, however, of the existence of contradictions within the processes of developing Nature, is the ever-recurring opposition between the *old* decaying factors and the *new* arising factors at any given stage. The new tendencies win, but they do not have it all their own way, and the eventual outcome is a kind of synthesis. Lucretius was well aware of this strife between the old and the new:—

". . . omnia migrant,
omnia commutat natura et vertere cogit.
namque aliut putrescit et aevo debile languet,
porro aliut clarescit et e contemptibus exit."

(All things depart;
For Nature changes all, and forces all
To transmutation; lo, this moulders down,
Aslack with weary eld, and that, again,
Prospers in glory, issuing from contempt.)[2]

So it has been said "Dialectics holds that internal contradictions are inherent in all natural phenomena; the struggle between the old and the new, that which is dying and that which is being born, constitutes the internal content of the developmental process."[3] To take a concrete example, we might say that at certain stages of the world's development, when complex molecules were first becoming stable, and again when life was originating, the forces of repulsion constituted the old, and the forces of aggregation constituted the new. The forces of aggregation never win entirely, for in Nature though victories may be decisive they are never total; the new synthesis at the higher level embodies elements of both the warring sides at the lower level. This is the secret of all high levels of organisation. But the history of the world shows that the forces of aggregation do, *on the whole*, succeed in their tasks.

[1] Cf. the valuable book of A. I. Oparin, *The Origin of Life* (Moscow, 1936, 2nd edn. enlarged 1941; Eng. tr. by S. Morgulis, New York, 1938).
[2] *De Rerum Natura*, V, 830 ff.
[3] J. Stalin, *Dialectical and Historical Materialism* (Lawrence & Wishart, London, 1941, p. 9).

Now there is a certain fundamental affinity between organicism and the dialectic. As Mortimer Adler has pointed out,[1] entities in opposition to one another are parts of a whole which stands on the level of their ultimate synthesis. The syntheses at all the successive levels of being, resolving the successive contradictions, form a series of envelopes, for they each include the elements of the contradictions on the levels below them as a series of parts. Like so many things in nature, the successive syntheses form a dendritic continuum or hierarchy of wholes.[2]

Whitehead as the Philosopher of Evolution.

And so we come to consider Whitehead's own contributions from the biologist's point of view. Unlike so many philosophers he has always appreciated the structure of our world in its *succession* and its *envelopes*. Perhaps one of his most famous and influential passages was that in which he said: "Science is taking on a new aspect which is neither purely physical nor purely biological. It is becoming the study of organisms. Biology is the study of the larger organisms, whereas physics is the study of the smaller organisms."[3] And so, regarding envelopes: "In surveying nature, we must remember that there are not only basic organisms whose ingredients are merely aspects of eternal objects [i.e. the ultimate particles of physics, each of which is related to everything else in the universe by its bare co-existence]. There are also organisms of organisms. Suppose for the moment and for the sake of simplicity, we assume, without any evidence, that electrons and hydrogen nuclei are such basic organisms. Then the atoms and the molecules are organisms of a higher type, which also represent a compact definite organic unity. When we come to the larger aggregations of matter, the organic unity fades into the background. It appears to be but faint and elementary, it is there, but the pattern is vague and indecisive. It is a mere aggregation of effects. When we come to living beings, the definiteness of pattern is recovered and the organic again rises into prominence."[4] Elsewhere, Whitehead

---

[1] M. J. Adler, *Dialectics* (London, 1927), pp. 164 ff; cf. also Stalin, loc. cit., p. 6, "Dialectics regards Nature as a connected and integral whole, in which phenomena are *organically* connected with, dependent on, and determined by, each other" (italics mine).

[2] Cf. the work of J. H. Woodger, "The Concept of Organism, etc.," in Quart. Rev. Biol., 1930, **5**, 1, 438 and 1931, **6**, 178.

[3] S & MW, p. 150.

[4] S & MW, p. 161.

elaborates this at length.[1] "The universe," he says, "achieves its values by reason of its co-ordination into societies of societies, and into societies of societies of societies. Thus an army is a society of regiments, and regiments are societies of men, and men are societies of cells, of blood, and of bones, together with the dominant society of personal human experience; and cells are societies of smaller physical entities such as protons, and so on. All of these societies presuppose the circumambient space of social physical activity."

So also with the successions. According to Whitehead, Nature exhibits itself as exemplifying a philosophy of the evolution of organisms subject to determinate conditions. Surveying the levels of organisation, he writes:[2] "One conclusion is the diverse modes of functioning which are produced by diverse modes of organisation. The second is the aspect of continuity between the different modes. There are borderline cases, which bridge the gaps. Often these are unstable and pass quickly, but span of existence is merely relative to our habits of human life. For infra-molecular occurrences, a second is a vast period of time. A third conclusion is the difference in the aspects of Nature as we change the scale of observation. Each scale of observation presents us with average effects proper to that scale."

Here is how he speaks of the emergence of mind.[3] "In so far as conceptual mentality does not intervene, the grand patterns pervading the environment are passed on with the inherited modes of adjustment. Here we find the patterns of activity studied by the physicists and chemists. Mentality is merely latent in all these occasions as thus studied. In the case of inorganic Nature any sporadic flashes are inoperative so far as our powers of discernment are concerned. The lowest stages of effective mentality, controlled by the inheritance of physical pattern, involve the faint direction of emphasis by unconscious ideal aim. The various examples of the higher forms of life exhibit the variety of grades of effectiveness of mentality. In the social animals there is evidence of flashes of mentality in the past which have degenerated into physical habits. Finally, in the higher mammals and more particularly in mankind, we have clear evidence of mentality habitually effective. In our own experience, our knowledge consciously entertained and systematised can only mean such mentality, directly observed."

Turning to the borderline of metaphysics, it is interesting to note

[1] AOI, p. 264; P & R, pp. 115 ff.; MOT, pp. 31 ff.
[2] N & L, p. 73.
[3] N & L, p. 94.

that Whitehead goes so far as to say that "a thoroughgoing evolutionary philosophy is inconsistent with [mechanical] materialism. The aboriginal stuff, or material, from which a materialist philosophy starts is incapable of evolution. This material is itself the ultimate substance. Evolution, on the [mechanical] materialist theory, is reduced to the role of being another word for the description of the changes of the external relations between portions of matter. There is nothing to evolve, because one set of external relations is as good as any other set of external relations. There can merely be change, purposeless and unprogressive. But the whole point of the modern doctrine is the evolution of the complex organisms from the antecedent states of less complex organisms. The doctrine thus cries aloud for a conception of organism as fundamental for nature. It also requires an underlying activity—a substantial activity—expressing itself in individual embodiments, and evolving in achievements of organism. The organism is a unit of emergent value, a real fusion of the characters of eternal objects, emerging for its own sake."[1] If in this passage Whitehead speaks like Lloyd-Morgan, we shall see others in which he speaks like Marx. Little though the philosophers of organic evolutionary naturalism may have borrowed from one another, they march in the same ranks.

Elsewhere Whitehead explains why he ignores for the most part nineteenth-century idealism. It was, he says, too much divorced from the scientific outlook, yet at the same time it swallowed the scientific scheme in its entirety and then explained it away as being an idea in some ultimate mentality. He leaves open, however, a final decision on the metaphysical issue—"However you take it, the idealistic schools have conspicuously failed to connect, in any organic fashion, the fact of nature with their idealistic philosophies. So far as concerns what will be said in these lectures [Science and the Modern World], your ultimate outlook may be realistic or idealistic. My point is that a further stage of provisional realism is required in which the scientific scheme is recast, and founded upon the ultimate concept of organism."[2] While this failure to close the door definitely on idealism has endeared him to theologians such as Thornton,[3] many scientists have preferred the robuster materialism of the marxists. No marxist, however, could be more strongly opposed to mechanical materialism than Whitehead.

---

[1] S & MW, p. 157.
[2] S & MW, p. 93.
[3] L. Thornton, *The Incarnate Lord* (London, 1930).

"My aim," he says,[1] "... is briefly to point out how both Newton's and Hume's contributions are each, in their way, gravely defective. They are right as far as they go. But they omit those aspects of the universe as experienced, and of our modes of experiencing, which jointly lead to the more penetrating ways of understanding. In the recent situations at Washington, D.C., the Hume-Newton modes of thought can only discern a complex transition of sensa and an entangled locomotion of molecules, while the deepest intuition of the whole world discerns the President of the United States inaugurating a new chapter in the history of mankind. In such ways the Hume-Newton interpretation omits our intuitive modes of understanding." In other words, what the President does is relevant to events at an extremely high level of organisation, and the concomitant atomic happenings are not directly concerned, though they underlie, and are entirely presupposed by, all that goes on at that high level.

Whitehead as the Philosopher of Organism.

Whitehead proceeds to his famous attack on the notion of "simple location."[2] "To say that a bit of matter has simple location means that, in expressing its spatio-temporal relations, it is adequate to state that it is where it is, in a definite finite region of space, and throughout a definite finite duration of time, *apart from any essential reference of the relations of that bit of matter to other regions of space and to other durations of time*. Again, this concept of simple location is independent of the controversy between the absolutist and relativist views of space or time. So long as any theory of space, or of time, can give a meaning, either absolute or relative, to the idea of a definite region of space, and of a definite duration of time, the idea of simple location has a perfectly definite meaning. This idea is the very foundation of the seventeenth-century scheme of Nature. Apart from it, the scheme is incapable of expression. I shall argue that among the primary elements of nature as apprehended in our immediate experience, there is no element whatever which possesses this character of simple location. It does not follow, however, that the science of the seventeenth century was simply wrong. I hold that by a process of constructive abstraction we can arrive at abstractions which are the simply-located bits of material, and at other abstractions which are the minds included in the scientific scheme. Accordingly the real

[1] N & L, p. 26.   [2] S & MW, p. 84 (italics inserted).

error is an example of what I have termed the fallacy of misplaced concreteness."

To the biologist all this was extremely welcome. If for three hundred years he had been a "mechanist" following in the footsteps of Descartes and la Mettrie, it was not because he felt satisfied with the seventeenth-century statistical picture of the fortuitous concourse of particles, each with a momentarily defined exact position in space, but because there was no other scheme by the aid of which he could proceed with the causal analysis of biological phenomena. The difficulties rose, of course, to a wild crescendo in the science of embryology at the turn of the century. When experimental embryology was put on a firm foundation by Wilhelm Roux, it was supposed that all eggs showed what is called "mosaic" development, that is to say, they would, if injured or divided, produce a finished organism lacking precisely all that would have developed from those parts which had been destroyed or removed. About 1895, however, the discovery was made (and this is what has secured Hans Driesch's name in history, not what he wrote long afterwards) that in many eggs, at any rate, all kinds of interferences could be made without affecting at all the embryo resulting. Large pieces could be removed from the egg, several blastomeres could be taken away, or the blastomeres could be shuffled at will, and yet a normal, though small-sized, embryo would result. Any one monad in the original egg-cell, then, was capable of forming any part of the finished embryo. Driesch was quite right in proclaiming that this was beyond the powers of any machine such as man has ever constructed, but he soon left the straight and narrow path by insinuating his non-material entelechy into the works as the inevitable transcendent mechanic or driver. C. D. Broad's comment deserves to be better known: "If you want a mind that will construct its own organism, you may as well postulate God at once. If He cannot perform such a feat, it is hardly likely that what has been hidden from the wise and prudent will be revealed to entelechies."[1]

Looking at the matter to-day after the passage of forty years of research in experimental morphology, we realise that what these early workers were up against was a very general process in development which we now speak of as Determination.[2] The individual

---

[1] C. D. Broad, Proc. Aristot. Soc., 1919, **19**, 123.

[2] As the new concepts came in in embryology, the old apparent necessity for postulating non-material factors went out. The process was excellently described by a

cells of the very young organism are not strictly determined as to their fate in the finished product, and this determination comes about as development goes on, partly at least through the action of chemical substances, about which we already know a good deal.[1] But the important point is that these chemical substances (the Evocators and Organisers) do not act at random, but faithfully in accordance with that plan of the body which is decreed by the characters of the species, whether embodied in the nuclear chromosomes or perhaps in the cytoplasm of the egg, a plan the field properties of which have earned for it the name of Individuation Field. Hence the fate of a given monad, protein molecule, atomic group, or what have you, in the original egg, is a function of its position in the whole. And thus we have a typical instance of the way in which the concept of simple location is hopelessly inadequate to cope with the facts arising in biological studies. The reader may be referred to Whitehead's own writings for an account of why it is inadequate in physics also, but others have made similar approaches, for example Wo. Köhler, starting from psychology, with his theory of physical "Gestalten."[2] According to Whitehead, all the things in the world are to be conceived of as modifications of conditions within space-time, extending throughout its whole range, but having a central focal region, which is in common speech "where the thing is." In topographic analogy, such as thermodynamicians use, the influence of the thing grades off past successive contours, like the slopes of Fujiyama, in every direction. The connection of this idea with the sort of fact which we are always meeting in biology, namely phenomena of field character, is obvious, and to-day the concept of field is equally widespread and necessary in biology as in physics.

To this may be added the following. The abstraction of classical seventeenth-century science from the life sciences had the effect, wrote Whitehead,[3] of bringing it about that dynamics, physics, and chemistry were the disciplines which guided the gradual transition

philosopher, S. Alexander (in his *Space, Time and Deity*, vol. ii, p. 65). After discussing the phenomena of pluripotence in embryonic development, he says, "Is there anything in these facts which is inexplicable when the initial constellation is considered? Instead of straightway postulating an entelechy to act as a guide, it would seem to me more reasonable to note that a given stage of material complexity is characterised by such and such special features, and that these are part and parcel of the principle or plan of the new order of complex." Determination is an empirical concept congruent with the facts of embryonic development.

[1] Cf. my *Biochemistry and Morphogenesis* (Cambridge, 1942).
[2] Wo. Köhler, *Die physische Gestalten*, 1920. [3] S & MW, p. 60.

from the full common-sense notions of the sixteenth century to the concept of Nature suggested by modern speculative physics. This change of view, occupying four hundred years, may be characterised as the transition from Space and Matter as the fundamental notions, to Process conceived of as a complex of Activity with internal relations between its various factors.[1] The phrasing here is important in view of what has to be said below concerning the transition from the concepts of Form and Matter to those of Organisation and Energy. The older point of view abstracted from any long-continuing change and conceived of the full reality of Nature at a single instant. It abstracted from any temporal duration, and characterised the interrelations in Nature solely by the distribution of matter in space. For the modern view, process, activity, and change, are, as for the dialectical materialists, the matter of fact. "At *an* instant there is nothing. Each instant is only a way of grouping matters of fact. There is no Nature at an instant. All interrelations of matters of fact involve transition in their essence. All realisation involves implication in the creative advance."[2]

Whitehead and the History of Science.

It is extremely interesting that Whitehead in this century, and Engels in the last century, both selected an almost identical group of scientific advances which they felt to have been the deciding factors in necessitating the great transition from the Renaissance or Newtonian outlook in science to the modern, dialectical, or organic. The three major discoveries selected by Engels were these:—

> "The first was the proof of the transformation of energy. All the innumerable operative causes in nature, which until then had led a mysterious, inexplicable, existence as so-called 'forces'—mechanical force, heat, radiation, electricity, magnetism, chemical affinity, etc.—are now proved to be special forms, modes of existence of one and the same energy, i.e. motion. The unity of all motion in nature is no longer a philosophical assertion but a fact of natural science.
>
> "The second—chronologically earlier—discovery was that of the organic cell by Schleiden and Schwann, of the cell as a unit, out of the multiplication of which, and its differentiation, all organisms, except the very lowest, arise and develop.

[1] N & L, p. 45.   [2] N & L, p. 48 (italics inserted).

"But an essential gap still remained. If all multicellular organisms—plants as well as animals, including man—grow from a single cell according to the law of cell-division, whence, then, comes the infinite variety of these organisms? This question was answered by the third great discovery, the theory of evolution, which was first presented in connected form and substantiated by Darwin."[1]

Whitehead speaks of four rather than three great advances, the first two being the idea of a field of physical activity pervading all space, and of atomism.[2] In the seventies of the last century some of the great departments of physics, such as light and electromagnetism, were established on the basis of waves in a continuous medium. But other sciences, such as chemistry, were established on the basis of ultimate particles or atoms and their interactions. Whitehead includes the cell-"theory" in biology as another example of the atomistic basis. It was, he says, in some respects, more revolutionary than the atomism of Dalton, for it introduced the notion of organism into the world of minute beings. "There had been a tendency to treat the atom as an ultimate entity, capable only of external relations, but Pasteur showed the decisive importance of the idea of organism at the stage of infinitesimal magnitude." Whitehead's second group of two new ideas comprises the law of the conservation of energy, and the doctrine of evolution. In energy-transformations, permanence underlies change. In evolution, permanence abdicates and change takes its place. There is, therefore, in the world an aspect of permanence and an aspect of change. In modern physics, wrote Whitehead,[3] "mass becomes the name for a quantity of energy considered in relation to some of its dynamical effects. This train of thought leads to the notion of energy being fundamental, thus displacing matter from that position. But energy is merely the name for the quantitative aspect of the structure of happenings; in short, it depends on the notion of the functioning of an organism." And evolution is the evolution of organisms of ever increasing organisation. As for the dialectical contradiction between particles and waves, that has only in our own time been, at any rate partially, resolved, with the modern theories of wave-mechanics, quantum mechanics, etc., about which it is hardly fitting that a biologist should speak.

[1] Appendix B to *Ludwig Feuerbach and the outcome of classical German philosophy*.
[2] S & MW, pp. 143 ff.   [3] S & MW, p. 149.

In parenthesis, having mentioned the agreement between Whitehead and Engels on the history of science, it is interesting to find at various places in Whitehead's writings remarkable echoes of marxist thought. One conceives that these originate from the congruity that there is between the dialectical conception of Nature and the organic conception. For example, when discussing dualism, Whitehead says, "The Universe is many because it is wholly and completely to be analysed into many final actualities. It is one because of the universal immanence. There is thus a dualism in this contrast between the unity and multiplicity. Throughout the Universe there reigns the *union of opposites* which is the ground of dualism."[1] And in another place, "In the past human life was lived in a bullock cart; in the future it will be lived in an aeroplane; and this change of speed amounts to a difference in *quality*."[2] More important, there are some fine passages where Whitehead expounds the changeableness of scientific formulations; the additions, distinctions, and modifications which have to be introduced perpetually into them; and the complete inadequacy of formal logic for science.[3] "We are told by logicians that a proposition must either be true or false, and that there is no middle term. But in practice, we may know that a proposition expresses an important truth, but that it is subject to limitations and qualifications which at present remain undiscovered." Clashes between theories are no sign of the failure of science, they are dialectical contradictions out of which much better approximations to truth will later arise. "A clash of doctrines is not a disaster—it is an opportunity."[4] A contradiction may be a sign of defeat in formal logic, but in science it marks the first step towards a victory. A reliance on scholastic and undialectical logic, which has marked so much writing in biological theory (e.g. the later works of H. Driesch) has been the reason why few biologists have troubled about it.

The Naturalness of the Mental and the Spiritual.

Reference was made above to Marx's phrase about making materialism not "misanthropic." An admirably parallel passage is to be found in Whitehead.[5] "In the same way as Descartes introduced the tradition of thought which kept subsequent philosophy in some measure of contact with the scientific movement, so Leibnitz introduced the

[1] AOI, p. 245.   [2] S & MW, p. 142.   [3] S & MW, p. 262.
[4] S & MW, p. 266.   [5] S & MW, p. 225.

alternative tradition that the entities which are the ultimate actual things, are in some sense procedures of organisation.... Kant reflected the two traditions, one upon the other. Kant was a scientist, but the schools deriving from him have had but slight effect on the mentality of the scientific world. It should be the task of the philosophical schools of this century to bring together the two streams into an expression of the world-picture derived from science, and thereby end the divorce of science from the affirmation of our aesthetic and ethical experiences." This, then, can only be done by recognising them for what they are, manifestations of the highest organisational levels, sublime indeed, but connected as surely with all the lower levels as the physical hands of a man playing a violin in an orchestra are with the claws of a crab.[1]

Such a connection involves what we have already had occasion to mention, the problem of the origin of mentality in evolution. There seems to be a bifurcation[2] here. As we ascend the organisational levels we seem to be led off in two separate directions, one the ascending series of social groups through animal associations to human community, the other the ascending series of stages of mental development. Perhaps it is not erroneous to regard the sociological and the psychological series as different aspects of one and the same set of high organisational levels. Only where the brain and central nervous system reaches its heights as in the primates does social organisation really develop, or conversely only where complexity is sufficient to allow of social life, intelligible communication and co-operative effort, does the mental life and its physical basis attain a high status.

The problem of "mind and matter" has always been the skeleton in biology's cupboard. Though generally abandoned to the philosophers, biologists never felt any satisfaction at the way in which their colleagues were dealing with it. When some twenty years ago the present writer constructed a chart to show the historical development of biochemistry and physiology since the fifteenth century, he built it around the mind-body problem. Later for a long time he thought that this had been a mistake, but perhaps it was really a correct and useful plan, though now it would require considerable revision. Physiology has had a curious history in this respect. Though the word was first used in its present sense by John Fernel in the

[1] Cf. Engels, *Dialectics of Nature* (Gesamtausgabe edn., Moscow, 1935), p. 695.
[2] I use the word not in its technical Whiteheadian sense.

sixteenth-century, the first textbook of physiology was the *De Homine* of Descartes, completed in 1637 but not published till 1662. Here mind and matter were absolutely separated. But as knowledge of the nervous system grew during the eighteenth and nineteenth centuries, it became more and more impossible to uphold this separation. As Whitehead acutely remarks,[1] "The effect of physiology was to put mind back into nature. The neurologist traces first the effect of stimuli along the bodily nerves, then integration at nerve centres, and finally the rise of a projective reference beyond the body with a resulting motor efficacy in renewed nervous excitement."

Elsewhere he sums up the situation thus:[2] "Descartes expresses dualism with the utmost distinctness. For him, there are material substances with spatial relations, and mental substances. The mental are external to the material substances. Neither type requires the other type for the completion of its essence. Their unexplained interrelations are unnecessary for their respective existences. In truth this formulation of the problem in terms of minds and matter is unfortunate; it omits the lower forms of life such as vegetation and the primitive animal types. These forms touch upon human mentality at their highest and upon inorganic nature at their lowest. The effect of this sharp distinction between nature and life has poisoned all subsequent philosophy. Even when the co-ordinate existence of the two types of actuality is abandoned, there is no proper fusion of the two in most modern schools of thought. For some, nature is mere appearance and mind the sole reality. For others, physical nature is the sole reality and mind an epiphenomenon. Here the phrases 'mere appearance' and 'epiphenomenon' obviously carry the implication of slight importance for the understanding of the final nature of things. The doctrine that I am maintaining is that neither physical nature nor life can be understood unless we fuse them together as essential factors in the composition of 'really real' things, whose interconnections and individual characters constitute the universe." This is a fine statement of the true scientific attitude to the problem of minds and bodies, and would be as acceptable to the dialectical materialists as to the emergent evolutionists. It means that when we speak of mind we mean (as Eddington would say) "mind (in the sense of Pavlov and Sherrington)" and not "mind (loud and prolonged applause)."

[1] S & MW, p. 213.　　　　　　　　　　[2] N & L, pp. 56, 57.

A Lack of Social Leadership.

Nearly all that has so far been said has been in praise of Whitehead's writings from the biological point of view, though with no attempt to pronounce upon the subtler points of his philosophy, a task impracticable for a working biologist. If any criticism were permitted, it would be that Whitehead has not been sufficiently outspoken in leading along the sociological and political directions in which his philosophy clearly points. It is true that he describes the creative aspect of evolution like any marxist, as the creation of their own environment by organisms. He says that here the single organism is helpless and that adequate action needs societies of co-operating organisms. But what tendencies in the world to-day are showing a capacity for such adequate action? If it were possible for Marx and Engels in the days of a capitalism comparatively mild and progressive to state their views uncompromisingly, whether right or wrong, about the line humanity must take towards higher levels of organisation, how much more necessary would it be in our own time, when the state power of fascism has arisen in a tottering social system, a power purporting falsely to be a higher level of organisation, but really no more than a mechanical tyranny. One looks in vain in Whitehead's writings for some clear lead among the social tendencies of our times. This is not to say that he has not sketched out, sometimes with brilliant detail, the historical origin of many of the features of economic individualism. Just as we made a connection earlier in this paper between economic individualism and seventeenth-century atomism, so Whitehead points out the connection between both these and the individualistic "cogito, ergo sum" of Descartes.[1] It led, he says, from private worlds of experience to private worlds of morals. Moreover, he suggests, not unconvincingly, that the assumption of the bare valuelessness of mere matter led to a lack of reverence in the treatment of natural and artistic beauty. The supreme ugliness of industrial civilisation, as it first arose, would thus be connected with its utter failure to recognise the unity of mind and matter at all the levels of organisation. But this is not what we are looking for. Whitehead's apparent inability to give a lead in his own time comes out especially strikingly in *Adventures of Ideas*, where the adventure of civilisation is discussed.[2] It is too abstract. It does not interlock with the concrete realities of political life. The objection or the defence

---

[1] S & MW, pp. 279 ff.   [2] AOI, pp. 352 ff.

that philosophers are not to be expected to descend into the arena of political struggles has no force. Philosophers themselves have said that the world would not be well until philosophers became kings. To-day kingship lies open to whoever cares to take it. It is said that Stalin was once asked when and where did Lenin expound dialectical materialism? He answered, "When and where did he not expound it?"—a curiously Confucian reply.

"The Rival Errors recognise their Loves."

But if the cobbler must stick to his last, the general upshot of this contribution is the great debt which biologists owe to A. N. Whitehead as the greatest living philosopher of organism. The epigram of old John Scott Haldane, neo-vitalist though he was, is coming true: "If physics and biology one day meet, and one of the two is swallowed up, that one will not be biology." In justice we should add that though it might perhaps be classical physics, it will not be physics itself, either; the two disciplines constituting indeed a Hegelian-Marxist contradiction of which the philosophy of organism is the synthesis. In conclusion it may be of interest to give two examples, from the author's own field, of the way in which the newer attitude is changing previous conceptions.

The question of the reducibility or irreducibility of biological facts to physico-chemical facts has already arisen several times above. Once the idea of a series of organic levels is reached, what we have to do is to seek to elucidate the regularities which occur at each of these levels without attempting either to force the higher or (anatomically) coarser processes into the framework of the lower or finer processes, or conversely to explain the lower by the higher. From this viewpoint the regularities discovered by experimental morphology will always have their validity; they cannot be affected by anything which either biochemistry or psychology may in the future discover. The behaviour, for instance, of an embryonic eye-cup isolated into saline solution—its capacity for self-differentiation, fusion with another eye-cup, lens-induction, regulation, etc. will always remain the same however our knowledge of biochemistry or biophysics may advance. This is the reason why prediction is possible at levels which, strictly speaking, we do not "understand" at all, for example, genetics. But though the biological regularities, once well and truly established, may remain for ever irrefragable, they will, considered alone, remain for ever meaningless. Meaning can only be introduced into our knowledge of the world by the simultaneous investigation of all the levels of complexity and organisation. Only in this way can we hope to understand how one is connected with the others. Only by understanding how one is connected with the others can we hope to see the meaningful integration of the evolving world in which organisation has been achieving its ever new triumphs.

The second question is one which has deeply concerned the present writer in his aim to unify biochemistry and morphology, namely the ancient problem of Form and Matter. Though not frequently discussed by Whitehead, it is fundamental for the biologist. The setting of μορφή and εἶδος against ὕλη by Aristotle has had an incalculably great influence on the historical development of biology. In the characteristic Greek art, sculpture, the form was certainly much more relevant to human interests than the marble or the bronze manifesting it. So for many centuries biologists devoted themselves to the study of animal form without much consideration of the matter with which it is indissolubly connected. It is not surprising that the numerous devils of vitalism found a congenial abode in the empty mansions of form thus suitably swept and garnished. The morphological tradition (originating perhaps from the idea of change as the privation of one form and the donation of another) was to think of matter far too simply, ignoring what we now know to be the vast complexity of chemical structures, and the unbroken line of sizes reaching from the sub-atomic levels to the particles of virus molecule size. Only in the light of the conception of organic levels can the saecular gulf between morphology and chemistry be bridged.

It is true that Aristotle held that there could be form without matter, though no matter without form. But according to him, the only entities which possessed form without matter, were the divine prime mover, the intelligent demiurges that moved the spheres, and perhaps the rational soul of man. Some of these are factors in which experimental science has never been very much interested. On the other hand, he maintained that there could be no matter without form, for however pure the matter was (even the chaotic primal menstrual matter which was the raw material of the embryo), it was always composed of the elements, i.e. always either hot or cold, wet or dry, and hence had a minimum of form. In its primitive way, this mirrors the position of modern science. Form is not the perquisite of the morphologist. It exists as the essential characteristic of the whole realm of organic chemistry, and cannot be excluded either from "inorganic" chemistry, or from nuclear physics. But at that level it blends without distinction into order as such, and hence we should do well to give up all the old arguments about Form and Matter, replacing these factors with two others more in accordance with modern knowledge of the universe: Organisation and Energy. From this point of view there can no longer be any barrier between morphology and chemistry. We may hope that the future will show us, not only what laws the form of living organisms exhibits at its own level, but also how these laws are integrated with those which appear at lower levels of organisation. This formulation is surely in line with Whitehead's philosophy of organism, and no less so with that of the dialectical materialists and the emergent evolutionists.[1]

Let us say once more then that in Whitehead's philosophy biologists find a view of the world which they are particularly well fitted to appreciate. Though dialectical materialism and emergent evolutionism have also much to teach us, they see in him the greatest and subtlest exponent of organic mechanism. These words are written in the College of Francis Glisson and William Harvey, of W. B. Hardy and Charles Sherrington. Isaac Newton's rooms, in that College of the Holy and Undivided Trinity, to which A. N. Whitehead himself belongs, lie only a stone's-throw away. From the neighbouring biological stronghold a word of deep respect and salutation goes out to the repairer of the onesideness of that Newtonian system which in its time was so profoundly progressive, the instaurator of the organic conception of the many-levelled world.

---

[1] Cf. the opinion expressed by Engels on form and matter:

"The whole of organic nature proves without exception that form and matter are identical or inseparable. Morphological and physical, form and function, are mutually determined. The differentiation of form (in the cell) conditions the differentiation of substance in muscle, skin, bone, epithelium, etc., and the differentiation of substance reacts back again and conditions new form" (*Dialectics of Nature*, Gesamtaugsabe edn., Moscow, 1935, p. 623).

Evolution and Thermodynamics
(1941)

THE development of modern science has led to a curious divergence of world-views. For the astronomers and the physicists the world is, in popular words, continually "running down" to a state of dead inertness when heat has been uniformly distributed through it. For the biologists and sociologists, a part of the world, at any rate (and for us a very important part) is undergoing a progressive development in which an upward trend is seen, lower states of organisation being succeeded by higher states. For the ordinary man the contradiction, if such it is, is serious, because many physicists, in expounding the former of these principles, the second law of thermodynamics, employ the word "organisation" and say it is always decreasing. Is there a real contradiction here? If so, how can it be resolved?[1]

At the outset it must be recognised that there is no question of rejecting the second law of thermodynamics. It is the basis of all our engineering technique which gives mankind power in controlling natural processes. The only question is, what exactly does it mean?

The Meaning of the Second Law of Thermodynamics.

In the general language which scientific workers use every day we say simply that free energy, that is, energy capable of doing work, is constantly decreasing, and bound energy (entropy) is correspondingly increasing. This absolutely irreversible process accompanies every natural change, whether physical or chemical. In a series of linked changes, however, there may be local decreases of entropy, provided that over the whole system entropy increases. The irreversibility of the "degradation" of energy has been identified by Eddington[2] and many other writers with the basis of our knowledge of the one-

---

[1] In thinking over these problems I have had the invaluable help and counsel of a number of friends: Professor Bernal, Dr. R. E. D. Clark, Dr. Danielli, Professor Dingle, Professor Donnan, Sir Arthur Eddington, Dr. Guggenheim, Dr. Robin Hill, Dr. Neuberger, Professor Polanyi, Dr. Shih-Chang Shen, Dr. Waddington, and Mr. Lancelot Whyte. Should this survey chance to fall under their eye, they will recognise points which they themselves have emphasised, but the whole responsibility for the general line and conclusions must necessarily be borne by me.

[2] e.g. in *The Nature of the Physical World* (Cambridge, 1928).

way character of time. Only in reversed cinematograph films, but never in nature, does water flow uphill of itself, or the smoke and gases of an exploding bomb recompress themselves, like the djinn of fable, into the reassembling case with its explosive content.

This description approaches the simple examples which are given by all elementary expositions of statistical mechanics.[1] If a number of atoms are introduced at one corner of a room, they will in a short space of time be found equally distributed within it (assuming that their kinetic energy is great enough to overcome their mutual attractions). In other words, it is impossible for a gas in a vacuum to occupy anything less than the whole of the space available. In the same way, if "hot" molecules are introduced at one corner of the room, and "cold" molecules at another, their collisions will soon ensure that all the molecules have the same velocity and that the temperature of the room is uniformly warm.

The significance of the second law of thermodynamics is, therefore, that all particles, when left to themselves, tend to become disarranged with respect to one another. Now such a process is similar to the shuffling of a pack of cards, or the random distribution of a quantity of black and white balls when continuously shaken together. "Shuffling," wrote Eddington, "is the only thing nature can never undo." High probability, therefore, is associated with randomness, low probability with the opposite, whatever you like to call it—perhaps arrangement or order. Hence the definition that entropy is the sum of the logarithms of the probabilities of the "complexions" of the parts of a system. A complexion or micro-state is simply an assembly of particles having the same velocity, rotational energy, rotational axis, etc. The more complexions there are in the system the less disordered it is. The presence of many complexions having very high or very low energies different from those in their vicinity is the condition under which useful work can be obtained from the system. This relatively unusual or improbable state constitutes what the physicist calls "order" or "arrangement." The corresponding "disorder" is measured by the logarithm of the probability.

Thus in an isolated system the net increase of entropy implies a net decrease of "order" and a net increase of "disorder." In such an isolated system a decrease of entropy can only occur in one part provided it is over-compensated by a simultaneous and greater

[1] e.g. *A System of Physical Chemistry* by W. C. McC. Lewis, 3 vols., vol. iii (London 1919).

increase of entropy in another part. The two parts must, of course, be interlinked by some sort of action, e.g. radiation. The essential physical meaning of increase of entropy is a loss of power of spontaneous action. Two bodies at different temperatures automatically tend to come to a common temperature, but having done that they stop doing anything. It is the same with matter at different levels (subject to the law of gravity), electricity at different potentials, and so on. The tendency is always to come to a state of passivity, and whenever such a tendency takes effect, entropy increases. The "running down" of the world is therefore a drift towards a state of relative quiescence.

From the point of view of other sciences and of the human world-view in general, it is important to note exactly what words the physicist uses to describe his improbable order and his probable disorder. If we turn to the writings of the American mathematical physicist, Willard Gibbs, the first among the great founders of thermodynamics, we find that he only uses one "ordinary" word to describe high entropic states, namely "mixed-up-ness."[1] This does not occur in any of his published writings, but only as the title of a paper which he had intended to write, and which was found among a list of such titles among his papers after his death. The opposite of "mixed-up-ness" is separatedness.

Later on, however, the practice grew up among physicists and astronomers of using the term "organisation" for pre-entropic states. Eddington has been a protagonist of this use, as the following passages from his *Nature of the Physical World* show:

> "We have to appeal to the one outstanding law, the second law of thermodynamics, to put some sense into the world. It opens up a new province of knowledge, namely, the study of organisation; and it is in connection with organisation that a direction of time-flow and a distinction between doing and undoing appears for the first time."[2]

Or again:

> "Let us now consider in detail how a random element brings the irrevocable into the world. When a stone falls it acquires kinetic energy, and the amount of the energy is just that which would be required to lift the stone back to its original height.

[1] *Collected Scientific Papers of J. Willard Gibbs* (London, 1906), p. 418.
[2] *The Nature of the Physical World*, p. 67.

By suitable arrangements, the kinetic energy can be made to perform this task; for example, if the stone is tied to a string it can alternately fall and reascend like a pendulum. But if the stone hits an obstacle its kinetic energy is converted into heat-energy. There is still the same quantity of energy, but even if we could scrape it together and put it through an engine we could not lift the stone back with it. What has happened to make the energy no longer serviceable? Looking microscopically at the falling stone we see an enormous multitude of molecules moving downwards with equal and parallel velocities—an organised motion like the march of a regiment. We have to notice two things, the *energy*, and the *organisation* of the energy. To return to its original height the stone must preserve both of them.

"When the stone falls on a sufficiently elastic surface the motion may be reversed without destroying the organisation. Each molecule is turned backwards and the whole array retires in good order to the starting-point.

> 'The famous Duke of York
> With twenty thousand men,
> He marched them up to the top of the hill
> And marched them down again.'

"History is not made that way. But what usually happens at the impact is that the molecules suffer more or less random collisions and rebound in all directions. They no longer conspire to make progress in any one direction; they have lost their organisation. Afterwards they continue to collide with one another and keep changing their direction of motion, but they never again find a common purpose. Organisation cannot be brought about by continued shuffling. And so, although the energy remains quantitatively sufficient (apart from unavoidable leakage which we suppose made good), it cannot lift the stone back. To restore the stone we must supply extraneous energy which has the required amount of organisation."[1]

A similar use of the term "organisation" occurs at many other places in Eddington's writings.[2] In the above example, it seems to mean no more than a group of uniformly directed motions, and such

[1] *The Nature of the Physical World*, p. 70.
[2] e.g. *The Nature of the Physical World*, p. 104.

an order might conceivably be regarded as the most primitive form of organisation. But to the biologist there is a sharp contradiction between this use of the term, and hence the view that the organisation in the universe is perpetually decreasing; and his own use of it which is associated with the evolutionary process. Here, in a part of the universe, at any rate, organisation is always increasing.

The matter is brought to a head when we find a physical chemist[1] describing the laws of probability and the second law of thermodynamics as the "law of morpholysis." This is well calculated to astonish the evolutionary morphologist, and makes imperative the effort to clarify the terminological situation.

The Meaning of the Law of Evolution.

Modern biology is nothing if not evolutionary. There are now no reasonable grounds for doubt that during successive ages after the first appearance of life upon the earth it took up a succession of new forms, each more highly organised than the last. This is not gainsaid in any way by the existence of highly adaptive parasitism and retrogression in certain types of plants and animals, nor by the fact that a hundred disadvantageous mutants may have to be produced for every one which is of evolutionary value. It is surprising that the theory of biological organisation is still in such a backward state. Though there are few penetrating accounts of it in the literature, every biologist has a rough working idea of what he means by it.[2] Here one may perhaps say that as we rise in the evolutionary scale from the viruses and protozoa to the social primates, there is

(1) a rise in the number of parts and envelopes[3] of the organism and the complexity of their morphological forms and geometrical relations;
(2) a rise in the effectiveness of the control of their functions by the organism as a whole;
(3) a rise in the degree of independence of the organism from its environment, involving diversification and extension of range of the organism's activities;

[1] R. E. D. Clark in Evangelical Quarterly, 1937, 9, 128, and in School Science Review, 1939, 21, 831; 1940, 21, 1117; see also his book *The Universe and God* (London, 1939).
[2] Attention may be drawn to R. W. Gerard's thoughtful discussion "Organism, Society and Science" in Sci. Monthly, 1940, 50, 340, 403 and 530; and to a paper of A. H. Kamiat, Internat. Journ. Ethics, 1933, 43, 395.
[3] See above, p. 184.

(4) a rise in the effectiveness with which the individual organism carries out its purposes of survival and reproduction,[1] including the power of moulding its environment.

There is nothing vitalistic about these criteria. All the levels of biological organisation are higher than the physico-chemical level, hence it is only natural that regularities will be expected to occur in them which cannot be seen at any physico-chemical level. But this is not to say that biochemistry and biophysics are not the fundamental sciences of biology. A living organism is both a "patterned mixed-up-ness," and a "patterned separatedness." The mere fact of the aggregation of millions of cells together into a functioning metazoon necessitates the provision of efficient means of control of the whole. Hence the mysterious similarity between the view that we see when looking down a microscope at a transparent blood-vessel, and the view of Broadway from the top of a skyscraper. Hence the mysterious similarity between the nerve fibres and the pyloned wire striding across the countryside. The new walls built in and around bombed buildings in London seem like scar tissue growing in an animal's body. A partly built steel-frame building, with its maze of pipes visible, seems like a metal organism in which human beings are parasitic. We speak of the "saturation" of anti-aircraft defences just as we speak of the saturation of an enzyme by its substrate, and the filtering of tanks through a line of defences resembles diapedesis.[2] Those thinkers who apply biological analogies to human society and its products are as foolish as any who would conversely try to persuade us that there really *are* micro-telephone-operators within the coelentrate nerve-net. The point is that the works of organisation have a certain similarity at all levels of their operation.

Furthermore, it is in general true that the higher the level of biological organisation, the more independent of the environment the

[1] There is a sense, of course, in which an amoeba is as organised as a man in that it carries out all the functions of assimilation, metabolism, reproduction, etc., but the difference lies in the variety of conditions under which it can do so, and the kind of limitations on the type of life which it can lead. There is also a sense in which all those species of plants and animals which have succeeded in persisting through evolutionary change are equally successful. But this is not the only criterion of success. Merely to persist is certainly the *sine qua non*, but we have also to consider under what variety of changed circumstances this persistence can occur, and also what the organism does with its persistence. One of the main objects in defining biological organisation and its rise during evolution is to get rid of as much subjectivism as possible in our outlook on other living things.
[2] C. M. Beadnell, Literary Guide, 1942, 57, 79.

organism is. Among the higher types decrease in number of reproductive products and corresponding increase of parental care illustrate this, and vast chapters of comparative physiology and biochemistry are devoted to the origin and development in evolution of body-temperature regulation, desiccation control, adjustment of the composition of the blood and the like. "The constancy of the internal medium," said Claude Bernard, in an aphorism popularised by Barcroft,[1] "is the condition of all free life." Nor is there in purposiveness anything that lies outside the scientific frame of reference. The Aristotelian theory of causation is irrelevant. Biological purposiveness and adaptation are concepts inseparable from biological facts, and as Donnan[2] has shown, there exist branches of mathematics, such as integro-differential equations, which may be able to cope with systems whose behaviour differs according to their past history. Consciousness is the highest phase of this behaviour and when the organisational level is reached at which psychological phenomena first appear, sociological phenomena appear too. One may say that biological organisation is as much an organisation of processes as of structures.

The point at issue is, then, whether the concepts of organisation as used by physicists and by biologists have anything in common.[3]

The Two Concepts of Organisation.

It is curious that this difficulty has not been more widely felt. Some thinkers have, indeed, been acutely troubled by it, for instance Rusk,[4] who spoke of a "conflict of currents," the physical world losing organisation and the biological world gaining it; and Ralph Lillie,[5] who opposed in an "apparent paradox" the "diversifying tendency" of evolution to the "dissipative tendency" making for

[1] "La Fixité du Milieu Intérieur est la Condition de la Vie libre," by J. Barcroft, Biol. Rev., 1932, 7, 24; and later as part of his book *Features in the Architecture of Physiological Function* (Cambridge, 1934).

[2] F. G. Donnan in Acta Biotheoretica, 1936, 2, 1.

[3] We consider mainly in this survey the rise in level of organisation during phylogenesis (evolution). But embryologists also see a rise in level of organisation during ontogenesis (the growth and differentiation of the individual). So far as we know, nothing that happens during embryonic development infringes the second law, and apart from the energy turnover in metabolic upkeep during development, there is little or no energy associated with the highly complex finished structure. For further information on this difficult subject see the section on energetics in *Chemical Embryology* (by J. Needham, Cambridge, 1931) and the review of O. Meyerhof (Handbuch d. Physik, 1926, 11, 243).

[4] R. D. Rusk, *Atoms, Man and Stars* (New York, 1937), p. 273.

[5] R. S. Lillie, Amer. Nat., 1934, 68, 318; Philos. of Sci., 1934, 1, 297.

uniformity, of the second law. So also Levy,[1] in his popular exposition of the sciences, remarks:

> "The fact is that the second law of thermodynamics, which regards systems as passing from orderly arrangement to disorderly randomness, classifies any future pattern or more complex orderly arrangement that may arise subsequent to the original order, as one of the innumerable accidental situations that have no special significance for man; as if a complex computing machine were indeed a random combination of parts. It may indeed mean that the energy of the original material from which the metal was drawn is now less available in one sense, but as a computing machine, it has now made available a mass of energy that was not previously capable of being tapped. Side by side, therefore, with the second law of the thermodynamics, in so far as it may be valid for large-scale systems—if it is so valid—there must exist a law for the evolution of novel forms of aggregated energy and the emergence of new qualities. A generalisation of this nature has not yet been made but that a general rule of this type must exist is evident."

One wonders why Levy did not allude at this point to the law of biological, psychological and sociological evolution. Long before, Engels, whom nothing escaped, had faced the problem, as we see from the following passage (in his *Ludwig Feuerbach*):

> "It is not necessary here to go into the question of whether this mode of outlook [evolutionary dialectical materialism] is thoroughly in accord with the present position of natural science which predicts a possible end for the earth, and for its habitability a fairly certain one; which therefore recognises that for the history of humanity also there is not only an ascending but also a descending curve. At any rate we still find ourselves a considerable distance from the turning point at which the historical course of society becomes one of descent, and we cannot expect Hegelian philosophy to have been concerned with a subject which natural science had at that time not as yet placed upon the agenda."[2]

[1] H. Levy, *Modern Science; A Study of Physical Science in the World Today* (London, 1939), p. 203.
[2] F. Engels, *Ludwig Feuerbach and the outcome of Classical German Philosophy* (London, n.d.), p. 22.

By this he would seem to have meant that a time may some day come when the struggle of mankind against the adverse conditions of life on our planet will have become so severe that further social evolution will become impossible. This was a sensible approach, but it was made in 1885, before statistical mechanics was fully developed, before the second law of thermodynamics had attained its present position of canonical importance, and before its interpreters had challenged the biologists by appropriating the term "organisation."

One reason why the apparent contradiction between the second law and the process of evolution has not caused more perplexity is that the attention of those few scientific thinkers who try to unify the world-view of science has been largely directed towards the question of the existence of disentropic phases within living matter itself. From the description of the second law already given it must have been quite clear that this generalisation has a statistical basis, and hence that if we had to deal with vessels so small that individual "complexions" could be separated, the statistical law valid for swarms of them might not in all cases hold good. Disentropic, "unusual," fluctuations might then, if amplified (and amplification is a process at which living matter is very efficient), account for such phenomena of high organisational level as "free will." Such a point of view has been ably put by Ralph Lillie[1] and discussed by Donnan.[2] It would be related to the standpoint of A. H. Compton,[3] and of G. N. Lewis,[4] who describes living organisms as "cheats in the game of entropy."

> "They alone," he wrote, "seem able to breast the great stream of apparently irreversible processes. These processes tear down, living things build up. While the rest of the world seems to move towards a dead level of uniformity, the living organism is evolving new substances and more and more intricate forms."[5]

Evidently he grasped the whole of the problem, but there is still one insuperable obstacle to the view that living organisms evade the second law of thermodynamics. It is simply that no evidence of any

[1] R. S. Lillie, Science, 1927, 66, 139; Journ. Philos., 1930, 27, 421; 1931, 28, 561; 1932, 29, 477; Amer. Naturalist, 1934, 68, 304; Philos. of Sci., 1934, 1, 296; 1937, 4, 202.
[2] F. G. Donnan, Journ. Gen. Physiol., 1927, 8, 685.
[3] A. H. Compton, *The Freedom of Man* (Yale, 1935).
[4] G. N. Lewis, *The Anatomy of Science* (Yale, 1926), ch. vi and viii.
[5] *The Anatomy of Science*, p. 178.

infringement of the second law on the part of living organisms has ever been forthcoming, and on the contrary a great deal of evidence that they obey it. Their life is always associated with, and depends upon, the processes of metabolism, in which there is always a net loss of free energy, complex organic compounds being broken down to CO_2 and water, and their energy being dispersed as irrecoverable heat. The processes of metabolism, moreover, are invariably very inefficient.[1]

The idea that living organisms might be cheats in the game of entropy originated from the conception of the so-called Clerk-Maxwell demon. Clerk-Maxwell, in his expositions of the second law, found it convenient to picture a vessel divided into two compartments which were connected by a hole and a trap-door which could be opened and shut at will by a "being whose faculties are so sharpened that he can follow every molecule in its course."[2] The demon could thus let through fast molecules but not slow ones, in which case, starting from a uniform temperature, one side would get hot and the other side cold. He could therefore easily, in Kelvin's words,[3] make water run uphill, one end of a poker red-hot and the other ice-cold, and sea-water fit to drink. The idea has thus been often put forward that living organisms evade the second law. "Es gibt seine Dämonen," cried Driesch,[4] "wir selbst sind sie." But as Clark[5] rightly says, what had really been proved was not that the second law was inapplicable to living matter, but that, *if* a mind could deal with individual molecules, and *if* it had suitable frictionless apparatus at its disposal, and *if* it was desirous of doing so at the moment when an observer happened to be looking, it *could* decrease entropy. These conditions are never, in practice, fulfilled. So far as we know, living organisms and their minds cannot handle molecules individually. The Maxwellian demon had in fact been endowed in its definition with just the qualities our minds possess of arranging, sorting and ordering. And our minds do not exist "in a vacuum," created from nothing; they belong to the highest stage in an evolutionary process continuous back to the most primitive single living cells. But the paradox is that in all this arranging, sorting and ordering which

[1] See J. Needham, "Recent Developments in the Philosophy of Biology," Quart. Rev. Biol., 1928, **3**, 77.
[2] J. Clerk-Maxwell, *Theory of Heat* (4th ed.), 1875, p. 328 ff.
[3] Kelvin, Proc. Roy. Inst., 1879, **9**, 113.
[4] Driesch, H., *Science and Philosophy of the Organism* (London, 1912), vol. ii, 202.
[5] Clark, R. E. D., School Sci. Rev., 1940, **21**, 1125.

living things perform, there is never any infringement of the second law of thermodynamics.

A Paradox with Social Significance.

At this point I wish to pause before pursuing the argument further in order to meet the criticism that the whole question is of purely academic interest. What difference does it make, one may ask, whether the world is thought to be "running down" or not, or whether the emergence of novelty in evolution is real? The answer is that our ideas on these questions have very marked and sometimes unrealised effects on our social behaviour.

The social significance of the first law of thermodynamics, i.e. the law of the conservation of energy, has long been realised. It was one of the basic pillars required for the development of industrial civilisation, as the physicists of the last century themselves knew very well. Crowther has summarised the matter:

> "The discovery of the conservation of energy is connected with the notion of exchange value. Capitalist civilisation cannot be operated without an exact knowledge of the equivalence of different forms of energy. ... When coal, electricity, gas and labour are to be sold in exchange, they must be measured and a common currency found for them. That currency is energy."[1]

And again:

> "All matter appeared to be made of electricity; industrial civilisation had at length succeeded in interpreting the universe in terms of one of its own concepts. The cosmos was conceived as made of one universal world material, electricity."[2]

Thus the thought introduced by the pre-Socratic Ionian philosophers 2500 years before, came to fruition. Thomson[3] has reminded us, in connection with the tyrant Midas and the first invention of gold coinage, that Heraclitus said, "Fire is the primary substance of which the world is made. Fire is exchanged for all things and all things for fire, just as goods for gold and gold for goods."

As with the first law, so with the second. Since it involves the time

---

[1] J. G. Crowther, *The Social Relations of Science* (London, 1941), p. 409.
[2] Loc. cit., p. 455.
[3] G. Thomson, *Aeschylus and Athens* (London, 1941), p. 85.

process more profoundly than any other scientific law, it immediately brings up all the "first and last things" of theology, from the creation to the last judgment. It has been a godsend to theologians filled with pessimism about human affairs, and delighted to find scientific backing for their despite of nature—"the heavens shall perish ... they shall all wax old as doth a garment." That arch-reactionary neo-Platonist, W. R. Inge, the former Dean of St. Paul's, devoted an entire book to demonstrating that since the universe is steadily approaching a state of thermal equilibrium and immobility, therefore (*a*) all evolution and human progress is an illusion and (*b*) men should return to what he calls the "philosophia perennis" of christianity, the conviction that all man's good lies in another life.[1]

> "We have here no continuing city, neither we ourselves nor the species to which we belong. Our citizenship is in heaven, in the eternal world to which even in this life we may ascend in heart and mind."[2]

> "So far as I can see," he goes on, "the purposes of God in history are finite, local, temporal, and for the most part individual. They all seem to point beyond themselves to the 'intelligible world' beyond the bourne of time and place.... In so far as the modern doctrine of the predestined progress of the species is only a spectral residuum of traditional eschatology, I think we must be prepared to surrender it."[3]

We might leave these stately passages to summarise Inge's position, were it not for another remark elsewhere in the same book which breathes the very spirit of the ecclesiastical department of the bourgeoisie:

> "Those who throw all their ideals into the future are as bankrupt as those who lent their money to the Russian or German governments during the war."[4]

---

[1] *God and the Astronomers* (London, 1933). Cf. also *The Fall of the Idols* (London, 1940). I confess that the general level of the argument in these books, though not its trend, reminds me of the well-known story in which a lecturer was giving a popular talk on the subject of the second law of thermodynamics and its implications. As soon as the meeting was thrown open for discussion, a member of the audience rose and said, "How long, sir, did you say it would be before the universe ran completely down?" The lecturer replied that he had said seven hundred million years. His questioner heaved a deep sigh of relief and said, "Thank God. I thought you had said seventy million years."

[2] Ibid., p. 137.   [3] Ibid., p. 172.   [4] Ibid., p. 28.

In justice to the theologians, it must be remembered that some have strongly contested Inge's views, notably Rashdall.[1]

Another reason for which theologians or theologically-minded scientists extol the second law is that it is tempting to identify the "winding-up" process, whatever it was, which started our galaxy off on its course with a maximum of free energy, with the act of creation by a personal deity. This always seems to me extremely premature, for we have no evidence that the universe does not operate in a cyclical way, periods of entropy-increase alternating with the appearance of new free energy. Milne[2] has pointed out that "the arrow of time" should more properly be regarded as a flight of arrows, since other galaxies are not at the same stage of their development as ours. Moreover, it may be that at the "edges" of the universe—if this expression means anything—free energy is continually being formed, so that like animals living in a stream or pipe which find the water always going by, we should see our world always "running down" but never reaching the end of the process. For those who like theological speculation, this might be regarded as a modern form of the doctrine of the "General Concourse" in which God must ever uphold the universe which he created. But I am unable to see that these speculations do anyone any good except those who are concerned to give ideological justification to ideas associated with backward social tendencies. It would be far better to await the further discoveries of astronomy and astro-physics with an open mind.

Thinkers approach the second law, therefore, with various forms of tacit bias, and these should be taken into account in considering what they say. One welcomes the degradation of energy and the disintegration of the world in the interests of other-worldly theology; another seeks evidence for a creator. I have no reason to suppose that I am without bias myself; in so far as anyone can state his own with any accuracy, I find the background of my thought to be the elucidation of the nature of life and man, the definition of the direction in which evolution has occurred, and the establishment of hope for man's struggles towards the perfect social order. The reason why the contradiction between the two concepts of organisation is so important is because the world-view, and hence the behaviour, of men in general is deeply affected by the "first and last things" of the world and of life and man within it.

[1] *Ideas and Ideals* (Oxford, 1928), by H. Rashdall.
[2] E. A. Milne, *Relativity, Gravitation and World Structure* (Oxford, 1935), p. 286.

The Two Views of the Contradiction.

We may now proceed to consider the two views which may be held about this contradiction. They are as follows:—

(1) The concepts of organisation as held by physicists and biologists are the same; but all biological, and hence social, organisation is kept going at the expense of an over-compensating degradation of energy in metabolic upkeep.

(2) The two concepts are quite different and incommensurable. We should distinguish between *Order* and *Organisation*.

First of all, can we find the first of these opinions explicitly stated?

Metabolism and Irreversibilty.

It is not difficult to do so. The physicist, Schrödinger, refers to the matter in his interesting book *Science and the Human Temperament*:

> "We are convinced," he writes, "that the second law governs all physical and chemical processes, even if they result in the most intricate and tangled phenomena, such as organic life, the genesis of a complicated world of organisms from primitive beginnings, and the rise and growth of human cultures. In this connection the physicist's belief in a continually increasing disorder seems somewhat paradoxical, and may easily lead to a very pessimistic misunderstanding of a thesis which actually implies nothing more than the specific meaning assigned to it by the physicist. Therefore a word of explanation is necessary.
>
> "We do not wish to assert anything more than that the total balance of disorder in nature is steadily on the increase. In individual sections of the universe, or in definite material systems, the movement may well be towards a higher degree of order, which is made possible because an adequate compensation occurs in some other systems. Now according to what the physicist calls 'order,' the heat stored up in the sun represents a fabulous provision for order, in so far as this heat has not yet been distributed equally over the whole universe (though its definite tendency is towards that dispersion) but is for the time being concentrated within a relatively small portion of space. The radiation of heat from the sun, of which a small proportion reaches us, is the compensating process making possible the manifold forms of life and movement on the earth, which fre-

quently present the features of increasing order. A small fraction of this tremendous dissipation suffices to maintain life on the earth by supplying the necessary amount of 'order,' but of course only so long as the prodigal parent, in its own frantically uneconomic way, is still able to afford the luxury of a planet which is decked out with cloud and wind, rushing rivers and foaming seas, and the gorgeous finery of flora and fauna and the striving millions of mankind."[1]

Thus in this charming passage we are to visualise biological order as identical with that order which thermodynamically is always disappearing, the local increase being more than compensated for by the decrease due to the cooling of the sun. Eddington, though more uncertainly, adopts a like view, in his *New Pathways in Science*.

"In using entropy as a signpost for time we must be careful to treat a properly isolated system. Isolation is necessary because a system can gain organisation by draining it from other contiguous systems. Evolution shows us that more highly organised systems develop as time goes on. This may be partly a question of definition, for it does not follow that organisation from the evolutionary point of view is to be reckoned according to the same measure as organisation from the entropy point of view. But in any case these highly developed systems may obtain their energy by a process of collection, not by creation. A human being as he grows from past to future becomes more and more highly organised—or so he fondly imagines. At first sight this appears to contradict the signpost law that the later instant corresponds to the greater disorganisation. But to apply the law we must make an isolated system of him. If we prevent him from acquiring organisation from external sources, if we cut off his consumption of food and drink and air, he will before long come to a state which everyone would recognise as a state of extreme 'disorganisation.' "[2]

Here, then, are statements of the view that there is no essential difference between thermodynamic and biological order. No doubt this is the simpler of the two alternatives.

Among the reflections which have led thinkers to support it is a

[1] E. Schrödinger, *Science and the Human Temperament* (London, 1935), p. 39.
[2] A. S. Eddington, *New Pathways in Science* (Cambridge, 1935), p. 56; see also *Nature*, 1931, **127**, 448.

recognition of the irreversibility which exists both in the second law and in biological evolution. The pioneer work of the palaeontologist Dollo[1] led to the generalisation which has since been called by his name. An organ which has been reduced in the course of evolutionary development never again reaches its original importance, and an organ which has altogether disappeared never again appears. Further, if in connection with adaptation to a new environment (such as aquatic, terrestrial, or aerial life) an organ is lost which was valuable in the previous environment, and if, as often happens, a secondary return to the previous environment occurs, this organ will not reappear. In its place some other organ will form a substitute. In a word, evolution is reversible in the sense that structures which have been gained can be lost, but it is irreversible in the sense that, once lost, these structures can never be regained. Since Dollo's time it has been shown that his generalisations hold good, not only for the morphological body structures which he elucidated, but also for numerous physiological and biochemical adaptations.[2]

In explanation, Dollo himself did not go much further than a vague appeal to the "indestructibility of the past." "In the last analysis," he wrote, "it is, like other natural laws, a question of probability. Evolution is a summation of determined individual variations in a determined order. For it to be reversible, there would have to be as many causes, acting in the inverse sense, as those which brought about the individual variations" (mutations, as we should say to-day) "which were the source of the prior transformations and their fixation. Such circumstances are too complex for us to suppose that they ever exist." This idea has something in common with the second law of thermodynamics. The universe is always passing from less probable to more probable states.

The position was further elaborated in the brilliant and unique book of Lotka, who defined evolution as the history of any system undergoing irreversible changes, and practically identified it with the second law. But although his discussion is one of the three or four greatest contributions to biological thought of the present century, he never really faces the problem that the second law involves a decrease, and the law of evolution an increase, of order. His pleasure

---

[1] L. Dollo, numerous papers, some of the more important of which are referred to in Biol. Rev., 1938, **13**, 225.

[2] By J. Needham, "Contributions of Chemical Physiology to the Problem of Reversibility in Evolution," Biol. Rev., 1938, **13**, 225.

at being able to unite all forms of irreversibility under one law leads him to an undue denigration of the genuine rise in level of organisation which evolution shows. Evolution, he rightly maintains, is not a mere changeful sequence. Mere unlikeness of two days does not tell us which preceded the other. It is necessary to know something of the character of this unlikeness.

> "In a vague way," he goes on, "this character is indicated by the term 'progress,' which is associated in popular conception with evolution. And the more rigorous scientific disciplines of biology, too, leave us with a not very clearly defined idea of 'progression' as one of the fundamental characteristics of those changes which are embraced by the term evolution. Such phrases as the 'passage from lower to higher forms,' which are often used to describe the direction of evolution, are vague, and undoubtedly contain an anthropomorphic element. At best they give every opportunity for divergence of opinion as to what constitutes a 'higher' form. If, on the other hand, it is stated that evolution proceeds from simpler to more complex forms, or from less specialised to more specialised forms, then the direction is but poorly defined, for the rule is at best one with many exceptions."[1]

And to this he adds in a footnote the remark of Bertrand Russell: "A process which led from amoeba to man appeared to the philosophers to be obviously a progress—though whether the amoeba would agree with this opinion is not known." After which he proceeds to take irreversibility as the principal character of the evolutionary process.

But this will not do. Denial of the rise in organisational level during evolution (and social evolution too) is not, and cannot be, acceptable to biologists. Definitions of what this means have been attempted above. Russell's wit is empty. It is mere verbiage (though it might in some circumstances be poetry) to talk about the opinions of molecules, or in that favourite phrase, "the hookworm's point of view" when its nervous system does not entitle it to have a point of view. Philosophers, on the contrary, are so entitled.

Here Max Planck has a relevant passage:

> "The second law of thermodynamics has frequently been applied outside physics. For example, attempts have been made

[1] A. J. Lotka, *Elements of Physical Biology* (Baltimore, 1925), p. 22.

to apply the principle that all physical events develop in one direction only, to biological evolution; a singularly unhappy attempt so long as the term evolution is associated with the idea of progress, perfection, or improvement. The principle of entropy is such that it can only deal with probabilities, and all that it really says is that a state improbable in itself, is followed on the average by a more probable state. Biologically interpreted, this principle points towards degeneration rather than improvement. The chaotic, the ordinary, and the common, is always more probable than the harmonious, the excellent, or the rare."[1]

And yet, in spite of these difficulties, and the general trend towards probable disorder, the rise in level of organisation during evolution has in fact occurred.

Patterns in the non-living world.

We come now to certain reflections which seem to suggest rather strongly that the thermodynamic principle of order is indeed fundamentally different from the biological principle of organisation. Biological organisation depends universally upon aggregations of particles, of high complexity. The molecule of protein, to say nothing of the paracrystalline protein micelle, is an entity so complex that though our analysis of it has begun, we are not as yet in sight of a clear understanding of it. Now although biological organisation, as we have seen, depends everywhere on a continuing metabolic upkeep, energy entering the plant as light or the animal as chemical energy in the foodstuffs, and being degraded in combustions and dissipated as heat; biological organisation is only the extrapolation of patterns already to be found in the non-living world. We cannot make any sharp line of distinction between the living and the non-living. At the level of the sub-microscopic viruses, they overlap. Some particles show some of the properties of life but not others. Some "dead" protein molecules are much bigger than the particles of some "living" viruses. Particles which show the properties of life can be had in paracrystalline, and even in crystalline form. There are many similarities between the morphology and behaviour of crystals and living organisms.[2]

Among the patterns found in the non-living world, crystalline

[1] M. Planck, *The Philosophy of Physics* (London, 1936), p. 101.
[2] Cf. such books as H. Przibram, *Die Anorganische Grenzgebiete d. Biologie* (Berlin, 1926), and F. Rinne, *Grenzfragen des Lebens* (Leipzig, 1931).

arrangement, and above all the arrangement of the more complex liquid crystals, is doubtless the most highly ordered and organised. But below the crystalline level there is the molecular level, and below that again the level of the atoms, some of which have a much more complex "solar system" of elementary physical particles than others. When crystals form spontaneously they do so in processes which involve decreases in free energy.[1] The physicist must therefore say that disorder has increased, but the biologist, as a student of patterns, cannot but say that there is more order and organisation in the well-arranged crystal than in its homogeneous mother-liquor or corresponding gas.[2]

So also in the development of our world. In its earliest stages, we are told, there were nothing but the elementary physical particles, and conditions were such that the atoms of the elements we know could not exist. But as the temperature of the earth grew colder, the atoms of the elements became stable and at last even the heaviest ones, with their dozens of revolving electrons, were able to persist.[3] Here, from the biologist's point of view, pattern and organisation had increased, but certainly from the physicist's point of view, order had decreased. The chaos which ensues upon the degradation of energy cannot therefore be the same chaos which existed at the beginning of the world before the atoms of the elements were stable. For that chaos coincided with a maximum of free energy, and the former accompanies its successive minima.

The point is, therefore, (a) that we cannot refuse to extend the concept of organisation downwards to include non-living patterned aggregations,[4] and (b) that since these require no continuing metabolic upkeep for their persistence, the "metabolic" theory which asserts

[1] Cf. G. N. Lewis & M. Randall, *Thermodynamics*, p. 122.
[2] Certain writers dispute this, e.g. R. O. Kapp (*Science versus Materialism*, London, 1940). They see in inorganic nature nothing but particles of different sorts "flying about" and "shaking down"; life alone introduces a plan or pattern. But may not life be the way the proteins shake down when in conjunction with lipoids, carbohydrates and certain other constituents?
[3] F. Wood-Jones (*Design and Purpose*, London, 1942, p. 55) has related how he, as a student, was struck by the similarity between Mendeléev's table of the elements and Huxley's table of biological types. Both have the status of evolutionary sequences. For an account of where Mendeléev's table of the elements stands to-day, see F. A. Paneth, Nature, 1942, **149**, 565.
[4] It should be noted that this conclusion is also that of certain philosophies; the emergent evolutionism of Lloyd-Morgan and others, the organic mechanism of Whitehead, and the dialectical materialism of Marx and Engels. It has also been finely expressed by the eminent physiologist Otto Meyerhof, in a review which is the best account of

that biological order and thermodynamic order are identical, but that the former is over-compensated, should be rejected in favour of a wider generalisation. We are led, in fact, to the second hypothesis mentioned above, namely that thermodynamic order and biological organisation are entirely different things.[1] Only we should now perhaps have to call the latter by some such term as "holistic" in order to indicate that pattern is not the biologist's perquisite, but occurs also at non-living levels.

As David Watson has well put it:

"The suggestion I wish to make is that all material organisation, whether living or lifeless, has its roots in the same facts, and that the symmetry and beauty of the products of the synthetic chemist, of geological formations, of trees, of flowers, and of young girls, are in essence traceable to the same kinds of designing agents."

Let us now return to the fundamental definition of Willard Gibbs —entropy is "mixed-up-ness." The opposite of mixed-up-ness is separatedness, not organisation. From this point of view, one can see that in the early stages of our world's development all the elementary particles of physics were in fact separated from one another, and free energy was then at its maximum. But as time went on, the temperature fell, and mixed-up-ness increased—or perhaps we ought to say that time went on because mixed-up-ness increased. The basic misconception we have unearthed comes to light here, namely that mixed-up-ness necessarily means chaotic mixed-up-ness; on the contrary there may also be patterned mixed-up-ness. Indeed it is hard to see how the most complex patterns could ever have been formed if there

---

thermodynamics and life from the physiological point of view (Handbuch d. Physik, 1926, **11**, 238). "In unseren Augen," he says, "stellen die Lebewesen eine höhere Organisationsform der unbelebten Materie dar, die sich etwa zur Organisation der Moleküle (oder Atome) so verhält, wie diese sich zu den Elektronen und Protonen, aus denen sie aufgebaut sind. Auch das Molekül und Atom stellt nur eine Struktur- und Funktionseinheit dar, von relativer Stabilität, von annähernd, aber nicht vollständig bestimmter Form, wechselndem Energiegehalt und numerisch bestimmten, aber individuell unbestimmten, d.h. austauschbaren, Elementarbestandteilen. Ja, darüber hinaus differieren sogar die Atome eines Elements unter sich unter gleichen Umständen, wie die Statistik des radioaktiven Zerfalls beweist, haben verschiedene Lebensdauer und besitzen demnach eine Individualität." Meyerhof, however, does not consider the problem of evolutionary rise in organisation as opposed to thermodynamic increase of disorder.

[1] This is also the standpoint of David L. Watson in an important paper (Quart. Rev. Biol., 1931, **6**, 143). Probability, he says, with Lotka, is essentially a matter of classification. An improbable event is one that is a member of a small class, and whether it is so or not depends on our system of classification. A classification based on morphological form and efficiency of function (means to ends) would give a different picture from that of statistical mechanics.

were not a number of different elements from which they could be formed. The world of life is a painting in full colours, not a monochrome. If this is accepted, we must make a sharp distinction between thermodynamic order or separatedness, and biological organisation or patterned mixed-up-ness. And the general upshot would be that the world has been moving steadily from a condition of universal separatedness (order) to one of general chaotic mixed-up-ness (thermodynamic disorder) plus local organisation (patterned mixed-up-ness).

The point could perhaps be illustrated by a homely analogy. Inside the nursery cupboard there are certain large boxes of bricks, each box containing bricks of identical colour and shape. When they are all tumbled out in confusion on the nursery floor we have the highly probable universe of the thermodynamician. But when in one corner of the room the bricks are assembled into a factory or a railway station, we have an analogy for the organising activities of life. There are two kinds of mixed-up-ness, indistinguishable for the physicist, but clearly visible to the biologist who is on the look-out for patterns.

The only thinker who seems to have arrived at somewhere approaching this position is the late J. S. Haldane.[1] He clearly expressed the idea that though with the passage of time thermodynamic mixed-up-ness constantly increases, this mixture does not necessarily give rise to chaotic states but on the contrary involves much pattern and organisation. Indeed, we may find the first traces of pattern even in those "fortuitous concourses" of particles which gases and homogeneous liquids were formerly thought to be, for it is now thought that temporary associations of particles exist in these systems, though they are of a duration so transient that they can only be observed by special methods. Such organisation would be around the molecular level.

Thermodynamic order and Biological organisation essentially different.

We have arrived, then, at the conclusion that thermodynamic order and biological organisation are two quite different things. But this is not to say that there are no connections between them. What Haldane said illustrates some of the connections. And there is another very simple way of showing how pattern may arise where there was

[1] See especially J. S. Haldane, Realist, 1930, 3, 10, and *The Philosophy of a Biologist*, (Oxford, 1936), p. 25.

none before, through the operation of the law of entropy. If we return to the example given earlier of two vessels filled with gas at different temperatures, and isolated from all other environment, we know, of course, that with the passage of time they will come to exact thermal equilibrium. When this point is reached it will certainly not be possible to get any further work out of the system, but a pattern has now appeared where it was not before. The system has passed from asymmetry to symmetry.[1]

It is, of course, a far cry from this simplest possible case of symmetry to the extraordinarily complex patterns of symmetry produced by living things, but it may be that this apparently jejune idea hides a profound truth. Every stage in the thermodynamic primrose path to the everlasting (but tepid) bonfire, has its own criteria of stability. In the earliest stages when free energy was maximal, even the atoms of the elements were not stable. Later on the minerals of which the earth's geography was built were stable for a long time before any living protoplasm was stable. The fibrous proteins (essential for the construction and maintenance of the higher forms of life[2]) can be regarded as "degenerated" linear polymers of the globular proteins, which themselves, as the work of the ultra-centrifuge school has shown, are polymers of a small and relatively simple unit. Polymerisation, too, is a notable feature of the carbohydrate group. Now polymerisation is a process which goes on with a decrease of free energy, just as crystallisation does, and since the most complex morphological forms require the most highly polymerised substances, Bernal[3] has pointed out that this fact helps to explain for us the contradiction between thermodynamic order and biological organisation.

From this point of view, life could be regarded as the characteristic stable form of the proteins.[4] "Life," wrote

[1] So also—as showing the different possible definitions of order—an equalitarian view might claim that when all the molecules had the same velocity the system had attained greatest order.
[2] Cf. W. T. Astbury, *Fundamentals of Fibre Structure* (Oxford, 1933), and J. Needham, *Order and Life* (Yale and Cambridge, 1936).
[3] J. D. Bernal, a paper to the Theoretical Biology Club, 1937.
[4] Stable, of course, only in a wide sense, including within itself a vast range of intermediate levels of stability. O. Meyerhof (Handbuch d. Physik, 1926, **11**, 240) gives an excellent description of the fundamental bases of life phenomena: "Ebenso wie die Atomkerne durch Einfangen von Elektronen Eigenschaften gewinnen, die die Elementarbestandteile noch nicht besitzen, die sich aber aus deren Eigenschaften herleiten, so würden die verschiedenen Lebensäusserungen—Wachstumsfähigkeit, Reizbarkeit, Stoffwechsel, Regeneration, usw.—aus der höheren Organisation der organischen, Molekülen entspringen, aber diese Moleküle selbst schon Eigenschaften enthalten, aus

Engels[1] in a famous definition, "is the mode of motion of the albuminous substances." So also in earlier essays, I was always deeply impressed by a fact so obvious that it never seemed to have occurred to many biological philosophers, namely, that proteins, carbohydrates and fats are never found in colloidal combination or even alone anywhere outside living organisms.[2] It was therefore irrelevant to demand of biologists that they should show the existence of similar phenomena in the inorganic world before applying the methods of physics and chemistry, as far as they would go, to the phenomena of life.

Goethe had faced this problem:

> "Wer will was lebendigs erkennen und beschreiben,
> Sucht erst den Geist herauszutreiben,
> Dann hat er die Teile in seiner Hand,
> Fehlt leider nur das geistige Band.
> '*Encheiresin Naturae*' nennt's die Chemie,
> Spottet ihrer Selbst, und weiss nicht wie."

which we might translate: "He who wishes to understand and describe living organisms first drives their spirit out of them, and then in his hands he finds nothing but the dead parts; the spiritual bond which united them has disappeared. Chemistry calls this 'the Manipulation of Nature,' insulting herself by so doing, and not even realising she does so." The investigation of this "spiritual bond" by the methods of natural science is the task of biology, and we have

---

denen jene im Fall der Organisierung abzuleiten sind. Der Bauplan der Lebewelt wäre daher schon in der unbelebten Natur vorgezeichnet. Neben der Eignung der Umwelt nach Temperatur, Feuchtigkeitsgrad, Kohlensäuregehalt, usw., ist vor allen die Elektronneutralität des Kohlenstoffs wesentlich, die die Bildung der homoiopolaren Verbindungen, das Aneinanderlagern langer und beliebig verzweigter Ketten von Radikalen veranlasst, womit die unendliche Mannigfaltigkeit der organischen Stoffe gegeben ist. So entstehen Moleküle von bedeutender Grösse und hinreichender Stabilität. Ihre kolloidalen Eigenschaften veranlassen widerum den gelatinösen Zustand der Zellen, der einen mittleren Grad des Diffusionsaustausche im Inneren ermöglicht; hiermit ist ein wichtiger Faktor der Reaktionsgeschwindigkeit der biologischen Umsetzungen gegeben, usw."

Cf. also R. S. Lillie's remark, "It seems clear that the concept of form and the concept of stability are closely allied. No form would be possible in chaos" (Philos. of Sci., 1937, 4, 220). Successive states of form in embryonic development are now treated from the point of view of stability levels (see my *Biochemistry and Morphogenesis*, pp. 112 ff.). Consider also an animal in a vessel isolated from any environment. If the energy present throughout the vessel were suddenly to be concentrated in the animal its temperature would rise to, say, 10,000° and it would die. Thermodynamic order would have increased and biological organisation would have decreased.

[1] F. Engels, *Anti-Dühring*, p. 94.
[2] Cf. SB, London, 1929, pp. 207 ff.

known too many victories in the past to fall a prey now to the pessimism of Mephistopheles.

It would be quite erroneous to say, therefore, that because thermodynamic order and biological organisation are two quite different things, there is any conflict between them. Only as the time-process goes on, only as the cosmic mixing proceeds, only as the temperature of the world cools; do the higher forms of aggregation, the higher patterns and levels of organisation, become possible and stable. The probability of their occurrence increases. The law of evolution is a kind of converse of the second law of thermodynamics, equally irreversible but contrary in tendency. We are reminded of the two components of Empedocles' world, φιλία, friendship, union, attraction; and νεῖκος, strife, dispersion, repulsion.

There remain one or two final points. First, we have not answered the question why a metabolic upkeep becomes necessary about the level of the beginnings of life. Basal metabolism, as it is called, that continuous slow rate of combustion which goes on all the time when the living organism is as much at rest as it can possibly be, is classically described as the energy turnover required to "keep the organism in being as a physical system," to maintain separation of phases, do work at membranes, etc., etc. It is fair to say that basal metabolism only appears slowly as we ascend the taxonomic scale. We hardly know as yet whether it is a concept which can be applied to bacteria, and there is much evidence that some forms of life, such as viruses, bacterial spores, and even some animal eggs, can remain for long periods of time in an inert state, consuming none of their stores and apparently as much without metabolism as any crystal. Presumably the higher living patterns do demand a good deal of additional energy to keep them going.

Secondly, there is the question of measurement. The nature of holistic organisation is certainly not susceptible of the same kind of measurement as thermodynamic mixed-up-ness, but we have no reason whatever for supposing that its measurement is impossible. When such a measurement has been achieved, it would be feasible to apply it also to human social evolution. I see no reason for doubting the possibility of this.

Thermodynamics and Social Evolution.

Mention of social evolution brings up a few reflections which inescapably present themselves. If the general principles here enunciated are sound, the most highly organised social communities should be the most stable, perhaps the most long-lasting. Here the social insects bear a testimony, for the forms of their different castes have been identified embedded in amber many thousands of years old, but they differ so severely from the primates in their morphology that they have less to teach us than is often supposed. The most highly organised social communities should also be the least wasteful. Hence it is significant that much of the criticism directed against our existing social order by those who wish to see a more highly organised state of society is precisely on the ground that our present arrangements are wasteful. Wasteful of human effort, when infinite care is devoted to the growing of a crop of coffee, only for it to be shovelled into locomotive fireboxes. Wasteful of energy, when heavy goods, the transport of which is not urgent, are transported by air or rail, while for purely financial reasons canals lie unused or derelict. Is there not a thermodynamic interpretation of justice? Is not injustice wasteful? Is not the failure to utilise to the maximum the available talent and genius of men a wasteful thing? As has been acutely said; "Those fundamental human rights which have so often been regarded as absolute postulates—liberty, equality, fraternity—turn out to be necessary conditions for a truly efficient productive mechanism and for the provision of sound social environment." That great aggregation of mankind to which we all look forward, the kingdom of heaven on earth, will be nothing if not efficient. There was a false contradiction in Richard Baxter's immortal remark: "I had rather go to heaven disorderly, than be damned in due order." The more truly orderly order is the more it approximates to the heavenly and this process is our own social evolution itself, our own history, in which it is our duty to participate, but we must always beware of mistaking the lesser forms of order for the greater, and, as Auden reminds us, we must be modest in our claims.

> "Great sedentary Caesars who
> Have pacified some dread tabu,
> Whose wits were able to withdraw
> The numen from some local law,
> And with a single concept brought
> Some ancient rubbish heap of thought
> To rational diversity;
> You are betrayed unless we see

> No Codex Gentium we make
> Is difficult for Truth to break.
> The Lex Abscondita evades
> The vigilantes in the glades;
> Now here, now there, one leaps and cries,
> 'I've got her and I claim the prize,'
> But when the rest catch up, he stands
> With just a torn blouse in his hands."

There is only one thing more to say. The increasing mixed-up-ness in the world gives the direction of time's arrow. Perhaps when it is possible to measure biological organisation the increasing patterned-ness will be found to lead to the same result. In the meantime it might be thought that what has happened to the world is that what was one single original pattern has split up into millions of subsidiary patterns—all were born from

> "The universe of pure extension where
> Only the universe itself was lonely. . . ."

The extraordinary thing is that Richard Baxter's contemporary, Thomas Browne, said it all in a flash of intuition in his pious seventeenth century way three hundred years ago, as if foreshadowing what we are thinking now:—

> "All things began in order, so shall they end, and so shall they begin again, according to the ordainer of order, and the mystical mathematicks of the city of heaven."

Integrative Levels: A Revaluation of the Idea of Progress

(Herbert Spencer Lecture at Oxford University, 1937)

> The flower of humanity, captive still in its germ, will blossom one day into the true form of man, like unto God, in a state of which no man on earth can imagine the greatness and the majesty.—HERDER.

> It is certain we shall attain
> No life till we stamp on all
> Life the tetragonal
> Pure symmetry of brain.—DAY LEWIS.

Statement of the Theme.

DISQUISITIONS without summaries are among the worst trials of the intellectual life. Only too often, on occasions such as the present one, when a scientist or a philosopher has the honour to stand before you, as now I have, forming one of a succession of plain thinkers who offer their conclusions for comment and criticism, he is content to leave his audience and his readers to gather his meaning as best they may without the assistance of any summary. In this case, however, a summary shall be provided, and instead of leaving it until the end, when your patience will have been, perhaps, severely taxed, it shall be given at the beginning in the form of a statement of the theme.

The theme of integrative levels is not one which we can approach without considerable hesitation, since the field which it covers is so wide and deep, no less than the whole nature of the world we know, and the way in which it has come into being. No one thinker can hope to do justice to this theme, and the only apology which may be offered for treating of it is that interest must always attach to what a specialist in any field of research may say when he abandons for a moment his speciality and looks boldly out to consider the world. Moreover, in taking the whole world for his province, your lecturer may the more easily, though a scientist, escape the condemnation of philosophers, who have always been rather interested in the world as a whole. The subject, then, to which our attention is to be given is the existence of levels of organisation[1] in the universe, successive forms of order in a scale of complexity and organisation.[2] This is a theme which that great man whose name we have in mind to-day, Herbert Spencer, the "synthetic philosopher," would at least have appreciated.[3] To-day it is no longer necessary, as it was in his time, to devote any effort to convincing people of the existence of evolutionary development in the world's history.[4] The cosmological changes which eventually produced a number of worlds, probably rather small among the galaxies, suitable for the existence of massed and complicated carbon compounds in the colloidal state, have become a commonplace background of our thought. So also the conception of biological evolution, in the course of which the many-celled animals and the plants arose from single-celled organisms probably somewhat resembling the autotrophic bacteria of to-day. A sharp change in organisational level often means that what were wholes on the lower level become parts on the new, e.g. protein crystals in cells, cells in metazoan organisms, and metazoan organisms in social units. Lastly, the anthropologists and ethnologists have familiarised all of us with the idea of evolutionary development in sociology, where we see the gradual development of human communities from the earliest beginnings of social relationships to the conception of the co-operative commonwealth now dawning upon the world.

But this great sweep of vision needs further elucidation. First, if we look carefully at the steps between the successive levels of organisa-

[1] I am not quite sure where the term "levels" was first used in this way, perhaps in S. Alexander's *Space, Time and Deity* (London, 1927, vol. ii, p. 52; 1st edn., 1920). This led to an interesting discussion among American authors (H. C. Brown, Journ. Philos., 1926, 23, 113; G. P. Conger, Journ. Philos., 1925, 22, 309) which I did not know about when this lecture was first written and printed. Nor did I know of the valuable book of the veteran American biologist, E. G. Conklin, *The Direction of Human Evolution* (New York, 1921), and that of the Manchester anatomist, F. Wood-Jones, *Design and Purpose* (London, 1942), which, broadly speaking, urge the same general viewpoint as that of the present book.

[2] Something approaching a definition of organisation will be given later, see p. 258.

[3] References to Herbert Spencer's own writings will be found in the footnotes as follows:

>FP, *First Principles* (6th edn., London, 1900).
>PB, *Principles of Biology* (London, 1898).
>PS, *Principles of Sociology* (London, 1876).
>A, *Autobiography* (London, 1904).

[4] Though it must be remembered that Roman Catholic writers are still fighting a rearguard action against it, and the devout engineer, R. O. Kapp, has attempted to reintroduce special creation under new terminology in his *Science versus Materialism* (London, 1940).

tion we find that the sharp lines of distinction are only made all the more sharp by the "mesoforms" which occur between them. Thus between living and non-living matter the realm of the crystalline represents the highest degree of organisation of which non-living matter is capable. It approaches, moreover, quite closely to the realm of the living in the phenomena presented by the so-called "liquid crystals," states of matter intermediate between the random orientation of a liquid and the almost absolute rigidity of the true crystal. These "paracrystals," with their internal structure and their directional properties, are closely related to living systems. Living systems, indeed, almost certainly contain many components of a paracrystalline nature. The viruses again, minute ultramicroscopic particles, probably represent some kind of intermediate form between living and lifeless.[1] But these forms of existence, the more clearly we understand them, will all the more clearly serve to bring out the essentially new elements of higher order which characterise the form of organisation we call life.

Secondly, there follows from the developmental nature of social organisation a conclusion which some thinkers, though otherwise clear-minded, have not been so ready to see, namely, that we have no reason to suppose that our present condition of civilisation is the last masterpiece of universal organisation, the highest form of order of which nature is capable. On the contrary, there are many grounds for seeing in collectivism a form of organisation as much above the *manière d'être* of middle-class nations as their form of order was superior to that of primitive tribes. It would hardly be going too far to say that the transition from economic individualism to the common ownership of the world's productive resources by humanity is a step similar in nature to the transition from lifeless proteins to the living cell, or from primitive savagery to the first community, so clear is the continuity between inorganic, biological, and social order. Thus, on such a view, the future state of social justice is seen to be no fantastic utopia, no desperate hope, but a form of organisation having the whole force of evolution behind it. But the acceptance of this implies a certain revaluation of the idea of progress. The idea of progress as applied to biological and social evolution fell into great discredit as the result of Victorian optimism. It was pointed out that evolution has often been regressive, that parasitism has been a widespread phenomenon in biology, and that before speaking of progress

[1] See the papers by N. W. Pirie and others in the Hopkins Presentation Volume, *Perspectives in Biochemistry* (Cambridge, 1937).

in evolution we should consider "the hookworm's point of view." Nevertheless, apart from the fact that the hookworm's nervous system does not entitle it to have a point of view, we cannot seriously bring ourselves to refuse to apply the concepts of higher and lower organisation to the animal world. Vertebrates *are*, in general, of higher organisation than invertebrates, mammals than other vertebrates, and human beings than other mammals. Again, in social affairs, the vast miseries caused by industrialisation and modern warfare were set against the doubtfully happier conditions of ancient times, and pessimistic conclusions adverse to the conception of progress were easily reached. But the time-scale was here insufficient, the exceedingly short space of time during which human civilisation has existed as compared with the time taken in biological evolution was forgotten. Post-Victorian pessimism mistook the development of a certain phase for the whole of progress itself. Of a famous Edwardian statesman it was said that he approached politics with the air of one who remembered that there had once been an ice age and that it was very likely there would be another. He was unnecessarily chilly. In the light of biology and sociology, those who remember that there were once autotrophic bacteria and that there will some day be a co-operative commonwealth of humanity, are better politicians.

So much for the theme of this disquisition. We shall naturally have to consider some aspects of Herbert Spencer's own thought as we develop its variations. But first it may perhaps be of interest if your lecturer takes leave to run over a few matters of personal interest, a few points on the intellectual travels which, in one form or another, it is everyone's fate to take. If one thing is more fundamental to the world-view outlined above than any other, it is the importance of the concept of Time.[1] And it was your lecturer's chance to become convinced of it in more than one major field of interest.

Time and the Theologians.

Perhaps exceptionally among students of science, he came to find theology, and especially the history of christian theology, one of the most fascinating of subjects. The intense persistence of so many minds, outstanding in their generations, to give rationality to the

---

[1] Cf. Samuel Alexander (*Space, Time and Deity*, London, 1927, vol. i, p. 36 fn.):

"I should say" (in contradistinction to Bertrand Russell) "that the importance of any particular time is rather practical than theoretical, but to realise the importance of Time as such is the gate of wisdom."

irrational and expression to the inexpressible, was an amazing phenomenon. The quarrels of theologians, orthodox and heretical alike, over words, letters, or even accents, in their formulations, were understandable enough, once certain premises were granted. The wonderful poetry of the liturgies carried symbolism to a point of daring surely hardly reached by any other great religion. Now it so happened that in the course of time he found himself influenced successively by two (still living) divines, W. R. Inge, the former Dean of St. Paul's, and Conrad Noel, for many years vicar of Thaxted in Essex.[1] It is true that they were but vehicles for the teaching of greater than they; Plotinus in the first case, Isaiah in the second. And it was after having experienced a profound attraction for the great tradition of christian mysticism that he came to feel, in the light of the prophetic and apocalyptic tradition, that the former was almost the evil genius of religion.

For the ancient Mediterranean thinkers, the world, which had neither beginning nor ending, was growing neither better nor worse. It has been powerfully argued (e.g. by Glover[2]) that the major contribution of christianity, and one of the principal reasons why it vanquished its competitors among the religions of the Roman empire, was precisely that it introduced change and hope into the stagnating sameness of the ancient world. But when asceticism, probably of Indian origin, outbalanced this new belief in the significance of time, the Neo-platonists, whether pagan or christian, had every reason they needed for turning away from the world and embarking on the ecstasies of the mystical contemplation of the One. "The intelligible world," writes Inge,[3] expounding Plotinus, "is timeless and spaceless, and contains the archetypes of the sensible world. The sensible world is our view of the intelligible world. When we say it does not exist, we mean that we shall not always see it in this form. The 'Ideas' are the ultimate form in which things are regarded by Intelligence, or God. $Νοῦς$ is described as at once $στάσις$ and $κίνησις$, that is, it is unchanging itself, but the whole cosmic process, which is ever in flux, is eternally

---

[1] Conrad Noel's books, *Byways of Belief* (London, 1912); *The Battle of the Flags* (London, 1922); *Life of Jesus* (London, 1937); and *Jesus the Heretic* (London, 1939), deserve to be even more widely known than they are, but his influence spread far and wide by the compellingness of his preaching and the exceptional beauty and grace of the Liturgy as celebrated at Thaxted.

[2] Glover, T. R., *The Conflict of Religions in the Early Roman Empire* (London, 1919).

[3] W. R. Inge, *Christian Mysticism* (London, 1921), p. 95; see also his *Philosophy of Plotinus* (London, 1929).

present to it as a process." A process, but not a progressive process. Where time brings no irreversible change, time is not important. It is strange how close to scientific thought theological thought has often been, for here we are reminded both of the seventeenth-century doctrine of the "general concourse," the upholding of the world by divine power without which everything would, it was thought, fly back into chaos again; and of the relations now understood between time and thermodynamic irreversibility. But for neo-platonic and hence christian mysticism, just as the sensible world is but a shadow of the intelligible, so action is a shadow of contemplation, suited only to weak-minded persons. This leads to what even Inge calls the heartless doctrine that to the wise man public calamities are only stage tragedies. It leads no less to the view that all such calamities are punishments for sin, since any action must be wrong. The mediaeval saint and visionary, Angela of Foligno, congratulated herself on the deaths of her mother, husband and children, "who were great obstacles in the way of God."

What a profound difference there is between this ascetic Graeco-Indian indifference to time, and the unsophisticated messianism which runs through most of the prophetic writings of the Hebrews, and on into the early Church, forming its other principal current. Here there is an intuition of time's irreversibility, the accomplishment of permanent gains, the belief in progressive change. Thus in Isaiah: "The voice of one that crieth, Prepare ye in the wilderness the way of the Lord, make straight in the desert a high way for our God. Every valley shall be exalted and every mountain and hill shall be made low; and the crooked shall be made straight and the rough places plain; and the glory of the Lord shall be revealed, and all flesh shall see it together: for the mouth of the Lord hath spoken it."[1] Or again: "The Lord God will come as a mighty one, and his arm shall rule for him; his reward is with him, and his recompense before him."[2] Or "Declare ye the former things, what they be, that we may consider them, and know the latter end of them; or show us things for to come."[3] Or Jeremiah: "And they shall come and sing in the height of Sion, and shall flow together unto the goodness of the Lord, for corn and wine and oil, and for the young of the flock and the herd; and their soul shall be as a watered garden, and they shall not sorrow any more at all. Then shall the girls rejoice in the dance, and the young men and the old together; for I will turn their mourning into

[1] Ch. 40, v. 3 ff.   [2] Ch. 40, v. 10.   [3] Ch. 41, v. 22.

joy, and will comfort them, and make them rejoice from their sorrow. ... There is hope for thy latter end, saith the Lord."[1]

Every word shows clearly a strong sense of the progressive time process. Things have been and shall be; they have been evil, it is promised that they shall be better.

After the period of the Gospels and in the early Church this attitude towards time became associated naturally with the conception of the Kingdom of God, *Regnum Dei*. In the later development of this conception, it came to mean, either the Church itself as a visible organisation, or an invisible company of the faithful, both of the dead and the living.[2] But the more your lecturer considered the history of the concept, the clearer it became that these ideas were later distortions, and that the primitive christians had held a much more materialist view of the *Regnum*, had thought of it rather as an earthly state of social justice which should, it was true, be brought in by the miraculous second coming of the Lord, but to which meanwhile all their own efforts should be tending. And in conformity with this "socialist" interpretation of the mind of the primitive church, he noted a number of facts to which as a rule little attention is given. Thus the communism of the Church of Jerusalem is generally, but inadequately, explained away by theological historians. Among the early theological movements, some of which were condemned as heresies or schisms, there are many traces of economic significance to be found, e.g. the *milites agonistici christi* of the North African Donatists, who seem to have been the shock-troops of an agrarian communist rebellion.[3] That there were elements of a hatred of communism in the mediaeval repressions of the Albigensians and Waldensians is more than probable.[4] All through the late middle ages, the peasant risings against their intolerable conditions were carried out in the name of christian comradeship, and often had the support of revolutionary clergy, as in the case of our English priest, John Ball. There are strong grounds for suspecting a social revolutionary element in the Lollards and the poor preachers of Wyclif.[5] And when it came to open warfare, the Anabaptists and the Taborites of the

[1] Ch. 31, vv. 2–20.
[2] For the historical development of the idea of the Kingdom, see Bp. A. Robertson's *Regnum Dei* (London, 1901).
[3] Cf. C. A. Scott, art. "Donatists" in Hastings' *Encyclopaedia of Religion and Ethics*.
[4] R. Pascal, "Communism in the Middle Ages and the Reformation" in *Christianity and the Social Revolution* (London, 1935).
[5] Cf. F. Engels, *The Peasant War in Germany* (New York, 1926).

sixteenth century[1] and the Levellers of the seventeenth[2] all adopted religious language and modes of thought. It is not to be suggested that any special significance attaches to this fact, for they had no other language or modes of thought at their disposal. But the essential point is that in all the best aspects of christianity, in those directions in which it has least turned its back upon human life and simple human happiness, the conviction of the reality of progress in time has been present. A time came when the old law gave place to the new. A time came when the people would suffer no longer the oppression of unchristian princes, but actually rose against them, and for a longer or shorter period withstood them. A time would eventually come when the Kingdom of God would be set up on earth, and a new world-order of love and comradeship would come into being.

With such words, we may seem to have travelled far from the cool consideration of biological and social evolution. But in fact we are not far removed from it, for there is a natural affinity between millenarism and evolutionary naturalism. Such primitive Christian ideas do justice to time and to progress, and in wandering, as a young man, from one theological realm to another, your lecturer came to see that the reality of time was fundamentally important. When he then returned to take up again the *locus classicus* of neo-platonic pessimism, the essay of W. R. Inge on progress,[3] it seemed to him, in spite of

[1] Cf. K. Kautsky, *Communism in Central Europe in the Time of the Reformation* (London 1897), and R. Pascal, *The Social Basis of the German Reformation* (London, 1933).
[2] Cf. E. Bernstein, *Cromwell and Communism* (London, 1930).
[3] "The Idea of Progress" in *Outspoken Essays*, vol. ii (London, 1923). It must with regret be recorded that some ten years afterwards Inge, in his Spencer Lecture for 1934, "Liberty and Natural Rights," gave his official blessing to fascism as the best form of human society yet devised. Opposition to the idea of progress is, indeed, a characteristic common to all fascist philosophers. Examples are easy to find. "The materialist outlook," wrote Otto Strasser, "has, as is well known, the idea of progress as one of its motive forces. There is no worse sort of fatalism than this spiritual hallucination that humanity has for millions of years now been marching along a road which leads for ever upward, decorated on the right and left with the milestones of development. How has this fixed idea become possible? Surely everyone knows from his own experience that life is a circle, not a line" (*Wir suchen Deutschland*, p. 165). So also Othmar Spann: "Darwin and Marx have done terrible harm to our civilisation by their mechanical" (*sic*) "conception of development. For this conception of development deprives every activity of value since today each one is overcome by tomorrow. And this has given birth to utilitarianism, materialism, and nihilism" (*sic*) "which are characteristic of our time" (*Kategorienlehre*, 1924, p. 211). And the Russian Orthodox Church adds its mite to the treasury. It has an Inge of its own. "The Humanism of the Renaissance," writes Berdyaev, "has not strengthened man but weakened him; that is the paradoxical *dénouement* of modern history. . . . European man strode into modern history full of confidence in himself and his creative powers, in this dawn everything seemed to depend upon his

all its learning, indescribably superficial. All it found to say about the eighteenth-century beliefs in reason and human perfectibility was that in France they "culminated in the delirium of the Terror." More than half a century after the beginnings of agricultural chemistry, and oblivious of modern methods of population control, it could still take seriously the views of Malthus. Neither history nor science, it concluded, give us any warrant for believing that humanity has advanced, except by accumulating knowledge and experience and the instruments of living; and the value of this social inheritance is "not beyond dispute." Nevertheless he retained sufficient admiration for a great, though perverse,[1] scholar, to continue to regard him as on a totally different level from Cro-Magnon man.

Time and the Biologists.

But side by side with these cogitations on the first and last things, your lecturer was occupied in his daily work with biology in general and biochemistry in particular. And since biochemistry is the most borderline of sciences, it was only natural that, like most reflective students of that subject, he should devote a good deal of attention to its philosophical position. That chemistry should indeed be able to cover the realms both of the inanimate and the animate, was in fact quite sufficiently a riddle in itself. The whole history of biochemistry, indeed, has been the scene of a persistent debate between those who have taken the hopeful view that the phenomena of life would one day be fully explicable in physico-chemical terms, and those who have thought themselves able to see in these phenomena evidences of some guiding influence—*spiritus rector*, *archaeus*, *vis formativa*, entelechy, or what you will—formally impossible to bring into relation with chemistry. Often enough these "vitalists," as they have been called, not content with prognostications of failure, have purported to give proofs of a more or less convincing nature, that the phenomena of life must ever resist scientific explanation.[2] During the first three decades of the present century the majority of working biologists and biochemists were not "vitalists" but "mechanists."

---

own creative powers, to which he put no frontiers or limits; today he leaves it to pass into an unknown epoch, discouraged, his faith in shreds, threatened with the loss for ever of the core of his personality" (*The End of Our Time*, 1933, p. 15). In 1940 Inge returned to his attack on the idea of progress in *The Fall of the Idols* (London).

[1] Another essay, *Our Present Discontents* (1919), will long remain a museum piece of upper middle-class spitefulness, unworthy of a christian, still less a priest.

[2] Such as Hans Driesch in his *Science and Philosophy of the Organism* (London, 1908).

About 1928 their position could fairly justly be summed up as follows: "Mechanists do not say that nothing is true or intelligible unless expressed in physico-chemical terms, they do not say that nothing takes place differently in living matter from what takes place in dead, they do not say that our present physics and chemistry are fully competent to explain the behaviour of living systems. What they do say is that the processes of living matter are subject to the same laws which govern the processes of dead matter, but that the laws operate in a more complicated medium; thus living things differ from dead things in degree and not in kind, and are, as it were, *extrapolations* from the inorganic."[1]

But the nature of this extrapolation was still obscure. The question entered a new phase, however, some ten years ago, with the publication of J. H. Woodger's remarkable book *Biological Principles*.[2] There it was laid down that the term "vitalism" should thenceforward be restricted to all propositions of the type "the living being consists of an X *in addition to* carbon, hydrogen, oxygen, nitrogen, etc. *plus organising relations*." Recognition of the objectivity and importance of organising relations had always been an empirical necessity, forced upon biologists by the very subject-matter of their science, but the issue was always confused by their inability to distinguish between the *organisation* of the living system and its supposed *anima*. With the abolition of souls and vital forces the genuine organising relations in the organism could become the object of scientific study. Before the contribution of Woodger, "organicism," as it had been called, had necessarily been of an obscurantist character,[3] since it was supposed as, for example, by J. S. Haldane, that the organising relations were themselves the *anima*, and as such inscrutable to scientific analysis.[4] To-day we are perfectly clear (though a few biologists may still fail to appreciate this point) that the organisation of living systems is the problem, not the axiomatic starting-point, of biological research. Organising relations exist, but they are not immune from scientific grasp and understanding. On the other hand, their laws are not

---

[1] SB, p. 247.   [2] (London, 1929).
[3] "Obscurantist" organicism was well castigated by N. I. Bukharin in the Marx Memorial Volume of the Moscow Academy of Science, 1933 (Eng. tr. *Marxism and Modern Thought*, p. 26).
[4] C. D. Broad in his *The Mind and its Place in Nature* (London, 1925), had argued along lines similar to Woodger's when he rejected both "substantial vitalism" and "biological mechanism" in favour of "emergent vitalism," but his treatment was for various reasons unsatisfactory and did not have much influence among biologists.

likely to be reducible[1] to the laws governing the behaviour of molecules at lower levels of complexity. It would be correct to say that the living differs from the dead in degree and not in kind because it is on a higher plane of complexity of organisation, but it would also be correct to say that it differs in kind since the laws of this higher organisation only operate there.

It may be of use to follow a little further the difference between the older dogmatic organicism and the new point of view. Organisation is inscrutable, it was urged, since any inorganic part instantly loses its relational properties on removal from the whole, and no means are available for rendering wholes transparent so that we can observe them while intact. But unfortunately these statements are not true. Woodger[2] has distinguished three main possibilities in the relation of organic part to organic whole: (*a*) independence, (*b*) functional dependence, (*c*) existential dependence. A part of the first sort would pursue its normal activities independently of whether it was in connection with its normal whole or not. A part of the second sort would be disorganised, if so isolated, and a part of the third sort would cease even to be recognisable. Dogmatic organicists, ignoring these distinctions, assumed that all parts are parts of the third sort. Yet this is certainly not the case. Liver cells synthesise glycogen and iris cells melanin in tissue culture as well as in the body. Isolated enzyme systems carry out their multifarious reactions in extracts as well as in the intact cells. Even existential dependence is a difficulty which can be overcome if means exist for making wholes "transparent," as by X-ray analysis of membrances or fibres, examination of living cells in polarising microscopes or ultra-violet spectrometers, or by "marking" in-going molecules by substituting isotope elements in them, such as heavy hydrogen or phosphorus.

It was a striking fact that in other countries other biologists had been coming to similar conclusions. In Russia, under the guidance of an elaborate philosophy at that time almost unknown here, a new organicism had been growing up, but so little were English men of science prepared for it that the very sensible and elaborate communications of the Russian delegation to the International Congress for the History of Science at the Science Museum at South Kensington,

---

[1] "Every new form of moving matter thus has its own special laws. But this enriched form and these new laws are not cut off by a Chinese wall from those historically preceding them. The latter still exist in 'sublated form' "; Bukharin, *loc. cit.*, p. 31.
[2] Proc. Aristot. Soc., 1932, 32, 117.

London, in 1931, were received with bewilderment.[1] "The true task of scientific research," said Zavadovsky,[2] "is not the violent identification of the biological and the physical, but the discovery of the qualitatively specific controlling principles which characterise the main features of every phenomenon, and the finding of methods of research appropriate to the phenomena studied. . . . It is necessary to renounce both the simplified reduction of some sciences to others, and also the sharp demarcations between the physical, biological, and socio-historical sciences." Again, in a passage which indicates a point of view closely similar to that already outlined, he writes, "Biological phenomena, historically connected with physical phenomena in inorganic nature, are none the less not only not reducible to physico-chemical or mechanical laws, but within their own limits as biological processes display different[3] and qualitatively distinct laws. But biological laws do not in the least lose thereby their material quality and cognisability, requiring only in each case methods of research appropriate to the phenomena studied." Or, in other words, biological order is both comprehensible and different from inorganic order. In France, similar views have been put forward, as, for instance, by Marcel Prenant,[4] also in accordance with the indications of materialist dialectics. This philosophy has been called the profoundest theory of natural evolution,[5] the theory of the nature of transformations and the origin of the qualitatively new,[6] indeed the natural methodology of science itself. It was striking to find that its conclusions upon a point of the most fundamental interest to the biochemist, the meaning of the transition from the dead to the living, should coincide with those which he had worked out independently by sincerely following the dictates of scientific common sense.

[1] English scholars owe a debt to Lancelot Hogben, who was one of the first about this time to try to translate dialectical materialism (more or less successfully) into English idiom; cf. his article in Psyche, 1931, **12**, 2. Certain mistakes afterwards pointed out (P. A. Sloan, Psyche, 1933, **13**, 178) do not diminish this debt. In the Aristotelian Society's Symposium on Materialism for 1928 there had been no mention of dialectical materialism, and a similar silence had reigned in the French symposium *Le Matérialisme Actuel* (Paris, 1920) to which H. Bergson, H. Poincaré, Ch. Gide, and others had contributed.

[2] Art. "The Physical and the Biological in the Process of Organic Evolution" in *Science at the Cross Roads* (Kniga, London, 1931).

[3] In the belief that the sense of the original is better conveyed, the word "different" is substituted for "varied" which actually appears in the text.

[4] M. Prenant, Bull. Soc. Philomath., Paris, 1933, **116**, 84.

[5] V. I. Lenin, "The Teachings of Karl Marx," in *Marx, Engels and Marxism* (London, 1931).

[6] J. D. Bernal, in *Aspects of Dialectical Materialism* (London, 1934), pp. 90 and 102.

The question had always been particularly serious for those biochemists who interested themselves in the problems of morphology. The enzymes involved in metabolism may be isolated and studied in relatively simple systems, analyses may be made of the substances entering and leaving the living body, even the blood and tissue fluids may be examined in relation to every conceivable bodily activity, change or disease—but all this avoids the main problem of biology, the origin, nature, and maintenance of specific organic structure. The building of a bridge between biochemical and morphological concepts is perhaps the most important task before biologists at the present time, and it may well be long before it is satisfactorily accomplished. But in the course of the present century several branches of study of great value in this connection have sprung up, particularly in embryology, where the changing organic form is the most obvious variable during development. Experimental and chemical embryology[1] together have made much progress towards the unification of chemical and morphological concepts. But this impressive change of morphological form takes place along the time-axis, and just as we have seen that in the far-away realm of the history of theology, the conviction of the importance of time was brought home to your lecturer, so also it was inescapable in the realm of biological science. In the development of the individual organism, as in that of organisms in general, progression took place from low to high complexity, from inferior

[1] It is interesting that Spencer himself had something to say on chemical embryology, in his time an almost uncharted field:

"The clearest, most numerous, and most varied illustrations of the advance in multiformity that accompanies the advance in integration, are furnished by living bodies. . . . The history of every plant and every animal, while it is a history of increasing bulk, is also a history of simultaneously-increasing differences among the parts. This transformation has several aspects. The chemical composition, which is almost uniform throughout the substance of a germ, vegetal or animal, gradually ceases to be uniform. The several compounds, nitrogenous and non-nitrogenous, which were homogeneously mixed, segregate by degrees, become diversely proportioned in diverse places, and produce new compounds by transformation or modification. . . . The yelk, or essential part of an animal-ovum, having components which are at first evenly diffused among one another, chemically transforms itself in like manner. Its proteid, its fats, its salts, become dissimilarly proportioned in different localities; and multiplication of isomeric forms leads to further mixtures and combinations that constitute minor distinctions of parts. Here a mass, darkening by accumulation of haematine, presently dissolves into blood. There fatty and albuminous matters uniting, compose nerve-tissue. At this spot the nitrogenous substance takes on the character of cartilage; at that calcareous salts, gathering together in the cartilage, lay the foundation of bone. All these chemical differentiations slowly become more marked and more numerous."
FP, p. 306.

to superior organisation. There had been a time when a certain level of organisation had not existed, there would come a time when far higher levels would appear. Time was the inevitable datum.

Time and Herbert Spencer.

We have now to give some consideration to the thought of the great "synthetic philosopher" himself. With much of what has so far been said, he would surely have been in definite agreement, since for him also the importance of the time-continuum, in which the irreversible world-process takes place, was cardinal.

"Evolution under its most general aspect," he wrote, is the integration of matter and the concomitant dissipation of motion; while Dissolution is the absorption of motion and concomitant disintegration of matter."[1] Sometimes the word integration as used by him seems to mean little more than a mere aggregation of undifferentiated matter,[2] but as soon as he comes to give examples of its function in evolution, we see that he means much what we mean when we speak of successive, and higher, levels of organisation. Total mass, he says, passes from a more diffused to a more consolidated state,[1] and the same process happens in every part that has a distinguishable individuality and finally there is an increase of combinations among such parts. Less coherence gives place to more coherence.[3] As his examples he takes, of course, the formation of solar systems from nebulae,[4] the development of the earth from a ball of hot gases,[5] the development of plants and animals in phylogeny and ontogeny,[6] and the rise of social relationships from primitive animal gregariousness to human communities of lower or higher order.[7] Then with perhaps some weakening of the imagination he goes on to say,[8] "Of the European nations, it may be further remarked, that in the tendency to form alliances, in the restraining influence exercised by governments over one another, in the system of settling international arrangements by congresses, as well as in the weakening of commercial barriers and the increasing facilities of communication, we see the beginnings of a European federation—a still larger integration than any now established." So throughout the range of levels, the same processes are seen—increase in the degree to which the parts constitute a cooperative assemblage, increase in the co-ordination of parts, increase

[1] FP, p. 261. [2] FP, pp. 258, 259. [3] FP, p. 299.
[4] FP, p. 281. [5] FP, p. 282. [6] FP, p. 284.
[7] FP, p. 288. [8] FP, p. 290.

in combination and juxtaposition and mutual dependence of the parts, and of the parts of the parts.[1]

Side by side with integration goes differentiation; the scission of wholes into parts, and parts into smaller parts.[2] Instances of this growing heterogeneity he finds in sidereal changes, in the changes of the earth's crust, in ontogenetic development (cf. the passage on chemical embryology already quoted) in phylogeny and in sociology. He ends by his celebrated definition of evolution;[3] evolution is a change from a relatively indefinite incoherent homogeneity to a relatively definite coherent heterogeneity, accompanied by integration of matter and concomitant dissipation of motion.[4] We may smile at what we suppose to be the presumption of such a cosmic formula, but we may find ourselves smiling on the wrong side of our faces, if, as is not unlikely, Herbert Spencer had hold of the right end of the stick. We should be foolish to put ourselves in the position of the devil, who was defined by the patristic writer, Hippolytus, as ὁ ἀντιτάττων τοῖς κοσμικοῖς, he who resists the world-process. With Spencer's attempt to elucidate the causes of this process, to say why evolution should go on at all (the instability of the homogeneous[5]), we need not here be concerned; the important point is his realisation of its universal scope. In reading his work to-day, we are likely to feel that he is most right where he emphasises integration and organisation rather than homogeneity and heterogeneity.

In Spencer's biological writings, too, there is much of great interest for the modern biologist who cares to know how ideas familiar to-day in science had their origin. The definition of life as the continuous adjustment of internal relations to external relations was his,[6] and so too was the conception of increasing independence of the environment accompanying increasing organisational level.[7] "One of those lowly gelatinous forms," he writes, "so transparent and colour-

[1] FP, p. 300.   [2] FP, pp. 301–14.   [3] FP, pp. 351 and 367.
[4] It is important to note that much of Spencer's argumentation depended on assumptions about energy which antedated modern statistical interpretations of the second law of thermodynamics. There is now, therefore, a certain contradiction here. The universe is passing, it is said, from less probable to more probable states, as if a basic shuffling process was continually at work. The word "organisation" is applied to the initial state of the universe, so that the increase of entropy must imply progressive disorganisation. The irreversibility of time is said to depend on this. We must, therefore, say either that thermodynamical organisation is quite a different thing from crystalline-biological-social organisation, or else that the persistent increase in the latter with time, which cannot be gainsaid, involves a correspondingly greater decrease of organisation somewhere else in the universe. See p. 207.
[5] FP, pp. 368 and 372.   [6] PB, I. 99.   [7] PB, I. 176.

less as to be with difficulty distinguished from the water it floats in, is not more like its medium in chemical, mechanical, optical, and thermal properties, than it is in the passivity with which it submits to all the influences and actions brought to bear upon it; while the mammal does not more widely differ from inanimate things in these properties, than it does in the activity with which it meets surrounding changes by compensating changes in itself." When in just twenty years' time we celebrate the centenary of Spencer's first formulation of this rule of increasing independence,[1] we shall be able to look back upon a vast structure of knowledge in comparative biochemistry and physiology which in many directions (e.g. osmotic regulation,[2] thermal regulation,[3] constancy of the internal medium,[4] respiratory pigments,[5] laws of nitrogen excretion,[6] etc.) has verified the synthetic philosopher's insight.

Nor was this much less remarkable in matters of embryology. His treatment of animal development as a passage from instability to stability has been profoundly justified in modern experimental embryology, in which the restriction of potentialities which goes on under the influence of the hierarchy of organiser-hormones bears him out.[7] Already before 1898 he had clearly enunciated the process we know now as "self-differentiation" under the name "autogenous development."[8] Even the organisation-centre, with its primary organiser-hormone, not discovered till 1924, he had adumbrated thirty years earlier, in the guise of an analogy with a party of colonists in new country, which forms for itself an organisation of "butty" or "boss" and those who work under his directions.[9]

But Spencer's treatment of sociological problems is of most interest for the present analysis. He has the great merit of having been among the first thinkers to apply evolutionary concepts to sociology, and for this we owe him a great debt.[10] Nevertheless we meet continually with the paradox that, having spoken so convincingly of the progressive integration of systems into ever higher levels of organisation, he

[1] It was first formulated in a review "Transcendental Physiology" in the Westminster Review in 1857 (A, I, 503). Whether his contemporary, Claude Bernard, who developed the concept of *fixité du milieu intérieur* had any hand in it, we do not know.
[2] See E. Baldwin, *Comparative Biochemistry* (Cambridge, 1937).
[3] See A. S. Pearse & F. G. Hall, *Homoiothermism* (New York, 1928).
[4] See J. Barcroft, *The Architecture of Physiological Function* (Cambridge, 1934).
[5] See A. C. Redfield, Quart. Rev. Biol., 1933, **8**, 31.
[6] See J. Needham, *Chemical Embryology* (Cambridge, 1931).
[7] FP, pp. 382 ff; see C. H. Waddington, *Organisers and Genes* (Cambridge, 1940).
[8] PB, I. 365.   [9] PB, I. 367.   [10] PS, I. 617.

stopped short at nineteenth-century England and found in its individualism nature's supreme achievement. The common ownership of the means of production, logical though it might be, did not seem to him the necessary next step in organisation, the next integrative level. There is thus a striking contradiction in his evolutionary thought. By what strange arguments was he able to convince himself that the liberal economic individualism of the mid-nineteenth century was the high state of integration to which all cosmic development had been tending? His life, his controversies with others, the internal evidence of his writings, may give us the clue.

A society, he says, is an organism.[1] How must we envisage its integration and differentiation? At once arises the question of the origin of classes and vocations, the division of labour.[2] It is the physiological division of labour, says Spencer, which makes the society, like the animal, a living whole. Complication of structure accompanies increase of mass, as the classes, military, priestly, slave, etc., differentiate—a progress from the general to the special. But he always fails to emphasise the different relationships of these classes to the production of goods or commodities, he always regards them with an exclusively political eye. Instead of seeking the origins of their economic relationships he elaborates, to a degree sometimes almost fantastic, the analogy between animal and social organisms. Thus the superior military class of warriors corresponds to the ectoderm, and the inferior class of cultivators, in close contact with the mechanism of food-supply, to the endoderm.[3] The origin of the State, he thinks, was the necessity of a centralised neural apparatus to co-ordinate the military activities of the organism-society against other societies. The more plausible explanation, that it was required as the instrument of domination of one class over the other, does not occur to him. As the peasants correspond to endoderm, so the king's council corresponds to medulla.[4]

In spite of this, however, Spencer was well aware of the limitations of the analogy.[5] There was, he said, a cardinal difference between the animal and the social aggregate. "In the one, consciousness is concentrated in a small part, in the other it is diffused throughout; all

[1] Though full of errors, both in fact and theory, the grandiose world history of O. Spengler, *The Decline of the West*, which since its first uncritical reception, has fallen into undeserved discredit, is strikingly in the Spencerian tradition, for it delineates the rise and fall of quite distinguishable cultural "organisms."
[2] PS, I. 468, 470, 491, 495.
[3] PS, I. 512.
[4] PS, I. 547, 552.
[5] PS, I. 479 and 612, A, I. 504.

the units possess the capacities for happiness and misery, if not in equal degrees, in degrees that approximate." The society exists for the benefit of its members, not its members for the benefit of the society. But there exists also another cardinal difference not mentioned by Spencer, namely, that once the early determinative processes of morphogenesis have gone on, all further cell-divisions produce like from like. Muscle-cells produce muscle-cells and neurons neurons. In a society every point corresponding to a cell-division means a completely new genetic shuffling of the pack of inheritable characters. Hereditary castes have thus no biological basis. Spencer shows some appreciation of this where he points out that succession by descent favours the maintenance of that which already exists, while succession by fitness favours transformations "and makes possible something better."[1]

It is fairly clear to-day that, if any form of society is most in accord with what we know of the biological basis of human common life, it is a *democracy that produces experts*. Consciousness, and all the higher human qualities, are dispersed pretty evenly throughout the world's population. But the unclearness of Spencer on this matter, it is interesting to note, together with a thoroughly uncritical social outlook, led in the hands of one of the most distinguished of your lecturer's predecessors, the great biologist William Bateson, to a striking Herbert Spencer lecture.[2] It well merits a short digression. Beginning badly by urging that biology must be the supreme guide in human affairs (as though sociology were not a higher organisational level than biology), it went on to say that, whereas "democracy regards class distinction as evil, we perceive it to be essential. . . . Maintenance of heterogeneity, of differentiation of members, is a condition of progress. The aim of social reform must be not to abolish class, but to provide that each individual shall so far as possible get into the right class and stay there, and usually his children after him."[3] By this mischievously misleading use of the term class, Bateson wholly surrendered the prestige of science into the hands of the middle-class employer and entrepreneur, assuring him that class-stratification was biologically sound and that the public-school tie covered all the best genes. Doubtless Bateson was referring to voca-

[1] PS, II. 260.
[2] The seventh; "Biological Fact and the Structure of Society" in *Herbert Spencer Lectures, Decennial Issue* (Oxford, 1916). The lecture was given in 1912.
[3] Loc. cit., pp. 31 and 32.

tional differences. He certainly underrated the genetic shuffling in each reproductive act. But to confuse the various vocations for which individuals should, of course, be as well suited as possible, with the division into classes differing according to their relation with the material means of production, some controlling these means, and others having access to them only by the grace and on the terms of the former, was a tragic mistake, worthy to stand side by side with the use of the theory of natural selection as a justification for *laissez-faire* economics.[1]

Now Spencer's main line of distinction in human societies was between "Predatory" and "Industrial."[2] The former type was one in which the army and the nation had a common structure, the army being the active manifestation of the nation. The latter type, though possessing some defence organisation, was characterised by voluntary co-operation in commercial transactions. It is clear that Spencer regarded the industrial type as higher than the predatory type. In describing examples of it (more or less convincing), his sympathies may be discerned; thus he speaks of "the amiable Bodo and Dhimals,"[3] "the industrious and peaceful Pueblos,"[4] and the development of free institutions in England.[5] In conformity with his view, already mentioned, on the good of the state as against that of the individual, he identifies the predatory organisation with the former and the industrial organisation with the latter. In this way we arrive at the classical position of nineteenth-century optimism, that all things work together for good for them that love profits, and that in an economic system where each man is for himself, the net resultant will always be for the benefit of all.

And now appears the remarkable, almost pathetic, naïveté of the synthetic philosopher. Spencer, approving of English capitalism in its

[1] On this a great deal could be written. Reference may be made to the following discussions as valuable starting-points for investigation:
C. Bouglé, art. "Darwinism and Sociology" in *Darwin and Modern Science*, ed. A. C. Seward (Cambridge, 1910); D. G. Ritchie, *Darwinism and Politics* (London 1889); J. G. Haycraft, *Darwinism and Race Progress* (London, 1900), L. Woltmann, *Die Darwinische Theorie u.d. Sozialismus* (Dusseldorf, 1899); O. Hertwig, *Zur Abwehr des ethischen, sozialen, und politischen Darwinismus* (Jena; 1921); J. S. Huxley, Proc. Brit. Assoc., 1936, p. 81.
We know now that the results of intra-specific competition are by no means necessarily good. As Huxley says, "they may be neutral, they may be a dangerous balance of useful and harmful, or they may be definitely deleterious."

[2] PS, I. 577, 590.
[3] PS, I. 585.
[4] PS, I. 585.
[5] PS, I. 587.

quiet home-transforming phase (the industrial type of society), viewed with horror the rise of British capitalism in its imperialistic phase.[1] It was, he said, the retrogression of the ideal form of society to a predatory phase. He abominated[2] "the recent growth of expenditure for army and navy, the making of fortifications, the formation of the volunteer force, the establishment of permanent camps, the repetition of autumn manœuvres, and the building of military stations throughout the kingdom." As his *Autobiography* shows,[3] he was even willing to take active part in the rather ineffective anti-militarist movements of his day. But the nature of his position forced him to fight on two separate fronts at the same time. He did not approve of imperialism, but neither did he approve of socialist, anti-individualist, legislation, favouring the working-class. By an extraordinary extension of the word "militarism" he was able to include both tendencies in the same condemnation. His individualism carried him to impressive lengths. Thus the compulsory notification of infectious diseases[1] and the unification of examinations for the learned professions[1] were regarded by him as unwarrantable interference with the freedom of the industrial social unit. Municipal housing,[4] nationalised telegraphs,[5] public museums,[5] even universal compulsory sanitary inspection and main drainage, were all put down as "tyrannical," "coercive philanthropy."[5] "Not by quick and certain penalty for breach of contract," he complained bitterly, "is adulteration to be remedied, but by public analysers."[5] Deeply ingrained in his sociology was the conception of free competition: "From the savings of the more worthy shall be taken by the tax-gatherer means of supplying the less worthy who have not saved."[5] Or again, in his autobiography we find a passage[6] in which, while describing one of his early essays, he says, "Among reasons given for reprobating the policy of guarding imprudent people against the dangers of reckless banking, one was that such a policy interferes with that normal process which brings benefit to the sagacious and disaster to the stupid." In such considerations men of Spencer's mind never stopped to reflect that the "less worthy" might also be the "more generous," or that "rapacious" might have been a better word in the sentence just quoted.[7]

[1] PS, I. 601. [2] PS, I. 602.
[3] A, II. 329 ff., 375 ff. Spencer attributed much of his breakdown in health to his activities in connection with a league for "anti-militancy" and "anti-aggression," which seems to have got little public support.
[4] PS, I. 604. [5] PS, I. 605. [6] A, II. 5.
[7] Spencer's discussion of communism illustrates this point strikingly. "State adminis-

We have already said that Spencer saw how succession by inheritance was a principle of social stability, while succession by fitness or efficiency was a principle of social efficiency.[1] Yet in his references to periods of social crisis when the principles of stability are challenged, when the forces tending towards a higher and hence more efficient level of organisation struggle openly with the forces of conservatism, he shows all the typical middle-class fear of supersession. He can even compare such upheavals with a gangrenous disease.[2] Just as in morbid changes, putrefactive dissolution may occur, so in "social changes of an abnormal kind, the disaffection initiating a political outbreak implies a loosening of the ties by which citizens are bound up into distinct classes and sub-classes. Agitation, growing into revolutionary meetings, fuses ranks that are usually separated. . . . When at last there comes positive insurrection, all magisterial and official powers, all class distinctions, all industrial differences, cease: organised society lapses into an unorganised aggregate of social units." A revolutionary might have reminded Spencer that not all dissolutions are morbid, that in the metamorphosis of insects, for instance, though there may be a histolysis, it is but the prelude to a new and more beautiful form of organisation.

Spencer and his Contemporaries.

The contradictions in Spencer's sociology appear again when we examine a few of the controversies and discussions in which he engaged. One of the most famous was that with the American sociologist Henry George. In Spencer's first book, *Social Statics*,[3] it was contended that the alienation of the land from the people at large had been inequitable, and that there should be a restoration of it to the State (the incorporated community) after compensation made to the existing landowners. "In later years," he wrote,[4] "I concluded

---

trations," he says (PS, II. 751), "worked by taxes falling in more than due proportion upon those whose greater powers have brought them greater means, will give to citizens of smaller powers more benefits than they have earned. And this burdening of the better by the worse, must check the evolution of a higher and more adapted nature." It is almost incredible that Spencer could have taken business success as his criterion of a high and adapted nature. "The diffusion," he says (loc. cit.), "of political power unaccompanied by the limitation of political functions, issues in communism. For the direct defrauding of the many by the few, it substitutes the indirect defrauding of the few by the many: evil proportionate to the inequity, being the result in the one case as in the other." An invitation to think out just what this means obviates any other comment.

[1] PS, II. 264.   [2] FP, p. 335.   [3] (London, 1850.)   [4] A, II. 459.

that a resumption on such terms would be a losing transaction, and that individual ownership under State-suzerainty ought to continue." George, who in his *Progress and Poverty*[1] had quite rightly advocated the nationalisation of the land, expecting it, however, to solve all social problems, replied with an attack (*A Perplexed Philosopher*)[2] which Spencer much resented. To George's accusation of consorting with Dukes, Spencer replied that it was only in a body partly founded by himself, the London Ratepayers' Defence League! It was a poor defence. Another controversy was with the Italian penologist Enrico Ferri,[3] who found that Spencer "stopped half-way in the logical consequences of his doctrine." In Ferri's view, natural selection and the struggle for existence in human society should not be interpreted as between individuals, but between classes. "Spencer believes," wrote Ferri in 1895 (loc. cit), "that universal evolution rules all orders of phenomena with the exception of the organisation of property, which he declares is destined to exist eternally in its individualistic form. Socialists, on the other hand, believe that it will itself also undergo a radical transformation . . . towards an increasing and complete socialisation of the means of production, which constitute the physical basis of social life and which ought not to, and will not, remain in the hands of a few individuals." Spencer complained bitterly in a letter to the Italian press.

With Beatrice Potter (later Mrs. Sidney Webb), the philosopher had close intellectual contact. While occupied with her long-continued studies on working-class conditions, she came to realise that the sphere of economics should include "social pathology," e.g. oppressive labour conditions. In 1886 she wrote to him putting this point as clearly as possible. He replied that on the contrary "political economy cannot recognise pathological states at all. If these states are due to the traversing of free competition and free contract which political economy assumes, the course of treatment is not the readjustment of the principles of political economy, but the re-establishment as far as possible of free competition and free contract." In other words, as she points out in her autobiography *My Apprenticeship*,[4] political economy is an account of the normal conditions in industry. But is not the first step to find out just what *are* the normal, or rather, the healthy, conditions in industry? Spencer, however, had made up his

[1] (London, 1881.)
[2] (London, 1893.)
[3] See E. Ferri's *Socialism and Positive Science* (London, 1906), p. 153.
[4] *My Apprenticeship*, by B. Webb (London, 1929), p. 292.

INTEGRATIVE LEVELS

mind *a priori* on this subject, and his flat refusal to question the dogma of free competition betrays, indeed, no little unconscious prejudice. The old man and the young woman agreed to go each on their way, and fifty years later the publication of the Webbs' great book on the Soviet Union[1] showed where the search for health in industrial relations had led.

That Spencer's sociology ended in a paradox has already been shown. It can hardly be understood except in the light of the thought of his contemporary, a man at least equally great, Karl Marx. Born within two years of each other and both living in England, they had, so far as we know, no contact of any kind.[2] Yet it is only in the light of the historical concepts of the great revolutionist in thought and action (as the Master of Balliol calls him[3]) that the failure of the great evolutionist to complete his edifice can be understood. Marx, with his friend Engels, was the genius, it has been said,[4] who continued and completed the three chief intellectual currents of the early nineteenth century, classical German philosophy, classical English political economy, and the French revolutionary doctrines which led to French socialism. Here we can only mention his combination of materialism with the dialectics of the Hegelian school, his economic formulations, especially that of surplus value, and his account of the roles of social classes in history. The former led to a philosophy, dialectical materialism, to which we have already had occasion to refer, which based

---

[1] *Soviet Communism*, by S. and B. Webb (London, 1935).

[2] In Spencer I find no reference to Marx; in the letters of Marx, however, there is one reference to Spencer. Though a little cruel, it is too amusing to omit. Writing to Engels on May 23, 1868, Marx says:

"Du scheinst mir auf dem Holzweg zu sein, mit Deiner Scheu, so einfache Figuren wie G-W-G etc. den englischen Revue-philister vorzuführen. Umgekehrt. Wenn du, wie ich, gezwungen gewesen wärst, die ökonomischen Artikel der Herren Lalor, H. Spencer, Macleod, etc. im Westminster Review, etc. zu lesen, so würdest Du sehn, dass alle die ökonomischen Trivialitäten so zum Hals dick haben—und auch wissen, dass ihre Leser sie dick haben—dass sie durch pseudophilosophical oder pseudoscientific slang die Schmiere zu würzen suchen. Der Pseudocharakter macht die Sache (die an sich = O) keineswegs leicht verständlich. Umgekehrt. Die Kunst besteht darin, den Leser so mystifizieren und ihm kopfbrechen so verursachen, damit er schliesslich zu seiner Beruhigung entdeckt, dass diese hard words nur Maskeraden von *loci communes* sind. Kommt hinzu, dass die Leser der Fortnightly wie der Westminster Review, sich smeicheln, die longest heads of England (der übrigen Welt, versteht sich von selbst) zu sein." Marx-Engels Gesamtausgabe, ed. Riazanov, Abt. III, Bd. 4, p. 58.

(G-W-G means Geld-Ware-Geld).

[3] In *Karl Marx's Capital* by A. D. Lindsay (Oxford 1935).

[4] By V. I. Lenin in *Marx, Engels and Marxism* (London, 1934), pp. 7 and 50.

itself upon that very evolutionary progression which Spencer described with so much care. His successive levels of integration are allowed for in the dialectics of nature, as in hardly any other philosophy. The concept of surplus value, says Dickinson,[1] has very unjustly shared in the logical discredit into which the labour theory of value has fallen. If Marx's theoretical foundation for it is unsatisfactory, some other must be found. But it is an undeniable fact of observation that the labour of men organised in society produces a surplus above the immediate requirements of the producers, and that this surplus may be disposed of in three main ways: (a) it may become the material basis of social growth, either absorbed in the support of an additional number of producers, or embodied in the increase and improvement of the means of production, or used to maintain more complex forms of social organisation, (b) it may appear as increased leisure or increased supply of consumption goods available for society as a whole, (c) it may be appropriated by a dominant class, appearing as rent, royalties, dividends, interest, profit, excessively high salaries, or various minor forms of privileged income. Here is a concept, the lack of which one deeply feels in reading Spencer's sociology. Again and again in his descriptions of the origins and nature of classes[2] he comes near to considering their relative economic *privileges*, but never clearly describes the phenomena of class-domination and the class-struggle. Hence he cannot realise the nature of the State; the neuro-muscular apparatus of control developed by the dominating class.

The history of all human society, past and present, wrote Marx and Engels in 1848, is the history of class-struggles.[3] "Freeman and slave, patrician and plebeian, baron and serf, guild-burgess and journeyman—in a word, oppressor and oppressed—stood in sharp opposition each to the other. They carried on perpetual warfare, sometimes masked, sometimes open and acknowledged; a warfare that invariably ended, either in a revolutionary change in the structure of society, or else in the common ruin of the contending classes." The history of the European West can only be understood in the light of this empirical fact. In the course of a long process extending over some four centuries, from about 1400 to 1800, the power of the feudal aristocracy gave place to the power of the middle-class. The

[1] H. D. Dickinson, Highway, 1936, p. 82.
[2] e.g. FP, p. 391.
[3] *Communist Manifesto*, 1848; first published in England, 1850.

work of historians, such as Pirenne[1] and Borkenau,[2] gives us an insight into the first embryonic origins of the new form of appropriation of surplus value which was later to be known as capitalism. Far back in the middle ages, the beginnings of long-distance transport, especially by sea, initiated the tradition of free finance and unlimited profit-making which did not come into its own until in seventeenth-century England the City of London, backing Cromwell's military force with all its might, made our country safe for Spencer's "sagacious" bankers. The process so brilliantly begun came to its fullness a hundred years later in France, when the chains of feudalism were finally broken and Europe's large-scale industries could develop in earnest. It must, of course, be emphasised that in those earlier days the middle-class "merchant venturers" and industrialists were the really progressive class. By Herbert Spencer's time, this was ceasing to be the case. Spencer stood just at the critical point when the middle-class was hesitating between the old policy of Manchester light industry and the new policy of Birmingham heavy industry. The export of finished goods was about to yield its hegemony to the direct exploitation of colonial countries and peoples, in a word, to imperialism. State expenditure upon the army was 14·9 million pounds in 1873–5, 18·1 in 1893–5, and 28·0 in 1911–13. Upon the navy it was 10·4 million pounds in 1873–5, 17·6 in 1893–5, and 45·3 in 1911–13. No wonder Spencer noted a "retrogression" from the industrial to the predatory state. The sociology for which he stood was that of the early nineteenth-century English middle class, favouring "cheap production and cheap government," i.e. low wages and no social legislation, a small army and navy, and even a moderate republicanism since bureaucracy and royalty might be thought unnecessary expenses. Still in the position of the early mill-owners and ironmasters, he objected equally to the expenses of imperialism and to the pressure towards social legislation exerted by the growing working-class movement. His grand sweep of vision from the nebulae to man truncated itself in the narrow prejudices of the dying class to which he belonged. But it is none the less valuable to us, for we can draw the conclusions he would not, and look forward to the inevitable further onward march of the principle of progressive integration and organisation.

[1] H. Pirenne, *Economic and Social History of Mediaeval Europe* (London, 1936).
[2] F. Borkenau, *Der Übergang von feudalen zum bürgerlichen Weltbild; Studien zur Geschichte der Philosophie der Manufakturperiode* (Paris, 1934).

The Giant Vista of Evolution.

Let us now take another look at the giant vista which has all along been the background of our thoughts. The stage once prepared by cosmic evolution for the appearance of life, what follows shows an ever-rising level of organisation.[1] The number of parts in the wholes increases, as also the complexity of their structure and their inter-relations, the centralisation and efficiency of the means of control (whether humoral or neural) and the flexibility and versatility of their actions on the external environment. The wholes become, indeed, ever more independent of the external environment; by regulation of exchanges in energy and materials an interior equilibrium is doggedly maintained, and though death destroys it in the individual, it continues in the species. If we run through any biological textbook,[2] we find abundant illustrations of this. Although some of the para-crystals already mentioned show a degree of complexity which seems to approach that of the simplest living organisms, it is the autotrophic bacteria which first exhibit the basic phenomena of the new level, reproduction and metabolism. They were (and are to-day) able to synthesise all the carbon compounds needed for their architecture from the carbon dioxide of the atmosphere by the aid of energy obtained from oxidations of inorganic substances (iron, sulphur, etc.). The many kinds of parasitic bacteria with which most of us are more familiar are to be supposed a regression from these primitive forms. But all was not regression, for by another big step cells grew enormously larger and the protozoa came into being. Some of these developed the photosynthetic mechanism, others did not. The former, when united together in colonies, became the first plants, the latter, similarly co-operating, became the first animals.[3] Then began that

[1] This "preparation of the stage" presents problems of much interest, the classical treatment of which is *The Fitness of the Environment* (New York, 1913), by Lawrence J. Henderson. Consideration of the properties of water, carbon dioxide, ammonia, etc., shows that if anything with properties at all akin to what we know as life was to develop, it must needs have the properties it actually did have. This reciprocal fitness of the environment greatly strengthens our view of the unity and continuity of the evolutionary process.

[2] For a student of another subject, an admirably philosophic introduction to biology as a whole can be had in the freshly-written book of H. H. Newman, *Outlines of General Zoology* (New York 1936). An excellent discussion of progress in evolution is given by J. S. Huxley in his presidential address to the British Association, 1936, pp. 96 ff.

[3] The beginnings of social behaviour, if not of social organisation, can be seen already in the aggregations of free-living protozoa (cf. H. S. Jennings, *Behaviour of the Lower Organisms*, New York, 1906; *The Beginnings of Social Behaviour in Unicellular Organisms*, Philadelphia, 1941; and in Science, 1941, **94**, 447; also W. C. Allee, *Animal Aggregations*, Chicago, 1931; Biol. Rev., 1934, **9**, 1; *The Social Life of Animals*, New York, 1938).

long procession of morphological forms and physiological achievements which the biologists have charted, with all its turning-points, the first coelomic organisation, the first endocrine mechanism, the first osmo-regulatory success, the first vertebral column, the first appearance of consciousness, the first making of a tool. At the point at which social life begins, factors set in so new as to constitute a recognisably higher level. Rational control of the environment now for the first time becomes a possibility.

The view of mind as a phenomenon of high organisational level, a quality of elaborate nervous organisation, is of course opposed by all idealist philosophers and many theologians, but it has wide support among psychologists and scientific philosophers. As examples I would mention a striking passage of the great psycho-pathologist, Henry Maudsley[1]; also the expressions of Samuel Alexander,[2] and the psychologists, R. G. Gordon,[3] C. K. Ogden,[4] E. B. Holt,[5] with many others. It need hardly be said that this view of mind has no connection with that which regards it as an "epiphenomenon." Perhaps few realise how well Lucretius stated the view of mind as a quality of high organisational levels in his great poem[6]:—

> "sed magni referre ea primum quantula constent,
> sensile quae faciunt, et qua sint praedita forma,
> motibus ordinibus posituris denique quae sint."
>
> (. . . but much it matters here
> Firstly, how small the seeds which thus compose
> The feeling thing, then, with what shapes endowed,
> And lastly what *positions* they assume
> What *motions*, what *arrangements*. . . .)

About the first beginnings of social organisation we know rather less than about some of the earlier, biological, stages. It is doubtful how far our consideration of humanity's problems can be assisted by a knowledge of the phenomena of social life in ants and bees (the social hymenoptera), for the anatomical nature of these animals, with its exoskeleton and rather inferior nervous system, is so far removed from our own.[7] The behaviour of the sub-human Primates has much

---

[1] *Body and Will* (London, 1883), p. 132.
[2] *Space, Time and Deity* (London, 1927), vol. i, p. xiii.
[3] *Personality* (London, 1926). [4] *The Meaning of Psychology* (New York, 1926).
[5] *The Concept of Consciousness* (London, 1914). [6] *De Rer. Nat.* II, 894.
[7] Popular writers such as J. Langdon-Davies in his *Short History of the Future* (London, 1936) go somewhat astray here.

more to tell us, but even that, as Zuckerman points out,[1] does not tell us much. Man's precursors probably lived a social life similar to that of all old-world monkeys and apes and were probably frugivorous. Probably at the beginning of the Pliocene, some twenty million years ago, when forests were reduced and the earth became more arid, a group of Primates with more plastic food-habits than the rest, managed to survive by becoming carnivorous. This transfer from a grazing to a hunting life must have had important social and sexual consequences, for with the change of diet there had to go a sexual division of labour in food-collection. Hence there had to be a repression of the dominant impulses which lead to polygyny in sub-human Primates. "The price of our emergence as Man," writes Zuckerman, "would seem to have been the overt renunciation of a dominant Primate impulse in the field of sex. The price of our continued existence may well be further repressions of dominant impulses, and *further developments*[2] of the co-operative behaviour whose beginnings can be vaguely seen in our transition from a simian to a human level of existence."

This was precisely Spencer's "blind spot." But we must take a closer look at co-operative behaviour, the necessary foundation for a higher order of human society. What has so far been said amounts to this, that evolution is not finished, that organisation has not yet reached its highest level, and that we can see the next stage in the co-operative commonwealth of humanity, the socialisation of the means of production.[3] Among the many evidences of this, there is space only to refer to two or three.

In the first place, the class-stratification such as we know it in all civilised communities, modelled on the pattern of Western Europe, is

[1] S. Zuckerman, article "The Biological Background of Social Behaviour," in *Further Papers in the Social Sciences* (London, 1937).

[2] Italics mine.

[3] For lack of space we pass here over the enormous gulf between the first beginnings of social organisation and the inadequacies of social organisation still existing in our own time. This emphasises the continuity of the factor of co-operation, with all the psychological adjustments that that implies. But it is essential to realise that within the sociological level there have been separate stages or levels, analogous perhaps to the mesoforms which we find as we pass from true liquids to true crystals. Thus it seems that the first attainment of an efficient agriculture was of enormous significance since it provided a food surplus and led to the formation of classes when this surplus came under partly or wholly private ownership. Similarly the first attainment of efficient machinery for the production of commodities had a profound effect, in the "industrial revolution" and the appearance of a truly proletarian class. The problems of today presuppose the accomplishment of these great changes in human social life; hence it is useless to ask why a classless society or the conscious control of production could not have been introduced centuries ago.

an ankylosis, a rigidity, a biological petrification, analogous in some ways to the armour-plating in which so many extinct animals spent their efforts. Jennings quite rightly says[1] that biology does not support democracy if democracy is defined as the belief that all human beings are alike or equal. It takes only common sense to see that they are each quite different from the other. But they all have needs and desires which could be satisfied and they all have contributions to make to the executive and productive power of the human collectivity ($\tau o\ \pi \lambda \acute{\eta} \rho \omega \mu a\ \tau \hat{\eta} s\ \text{'}E\kappa \kappa \lambda \eta \sigma \acute{\iota} a s$ as Chrysostom would have said). If democracy is defined as such a constitution of society that any part of the mass can in time supply individuals fitted for all its functions, then biology sanctions democracy. A democracy that produces experts.[2] Now it is painfully clear that in a class-stratified society there are very grave hindrances to this free utilisation of existing ability. There is no real equality of opportunity. Ninety per cent of the leaders are drawn from 15 per cent of the community. Geniuses and unusual types are likely to be stifled in childhood. There is a crushing effect on the very birth of initiative and constructive ability among the masses of the workers. "Not merely poverty and bad living-conditions, but soul-killing cap-touching subjection to a master class and the consciousness of toiling to produce profits for that class, deadens initiative and rouses hostility and antagonism to the whole industrial machine."[3]

But not only does society, in this lower stage of organisation, fail to draw upon anything like the full force of good gene-combinations that exist within it; it also fails to create as many of these as would otherwise exist. The whole rationale of the sexual reproductive system, from its beginning among the lower invertebrates upwards, can only be understood as a mechanism for producing an almost infinite diversity of qualities among the individuals of a species. Gene packs are shuffled anew in every reproductive act. The wider the range of individual differences the greater the chance of favourable variations. Yet in class-stratified human societies very severe checks are placed upon the mating-choices of individuals, a procedure quite irrational sociologically and ripe for conscious abolition.

The fact that the class-stratification has arisen, as we have said, from differences in the relations of men to tools and productive

[1] H. S. Jennings, *The Biological Basis of Human Nature* (London, 1931), p. 221.
[2] Jennings, loc. cit.
[3] *Britain without Capitalists* (London, 1936), p. 48.

resources, some being owners of these essential things, and others only having access to them on the owners' terms, also brought about a situation in which the accumulation of personal wealth is the only recognised sign of success. In Herbert Spencer's own thought we have had frequent occasion to remark upon it. The "sagacious," the "more worthy," the "prudent" etc., shall prosper, the weak shall go to the wall. Spencer was perfectly well aware that this psychological valuation might have biological consequences; he hoped it would. But the puzzle is that he should have been so certain that the characteristics which lead men to rise economically in a class-stratified capitalist society were those most desirable from a social point of view. Where predatory rather than co-operative behaviour wins the day, the path towards the higher social organisation is closed.[1] It was strange that Spencer could not see that the very predatoriness which he described in primitive societies of the "military" type saturated also even the most highly developed societies of what he called the "industrial" type. By a curious and happy irony, however, the effects of this high valuation of socially undesirable qualities have been much less than might have been the case, for the birth-rate of the socially successful groups has for long been far below that of the socially unsuccessful, and hence relatively uneducated, workers. The use of the expression, "survival of the fittest" for social success has therefore attained a definitely comic level. The only Darwinian meaning the term could have was as applied to those who send the largest number of offspring into the next generation. The successful capitalist, therefore, might be fittest for having a good time, but not for transmitting his genes to posterity.

The foregoing biological arguments have shown that the higher level of integration or organisation of the classless society would be greatly preferable to the class-stratified society. At this point the plain man might well object that a society with two, three, or four classes must surely be a more complex system than a classless one. To this, however, an obvious biological analogy provides the answer. We might as well assert that an annelid with twenty or thirty ganglia down its body is more complex than a mammal with a highly developed single brain. The almost unimaginable complexity of neurons, synapses, commissures, etc. in the human brain, forms a far higher organisational level than that of the annelid

---

[1] This is the theme of the classical essay of the geneticist, H. J. Muller, "The Dominance of Economics over Eugenics," Sci. Monthly, 1933, **37**, 40.

ganglia.[1] The human brain, indeed, is the outward and visible sign of the most fundamental of human characteristics, that by virtue of which sociology is a higher level than biology; the possession of consciousness. It follows that the more control consciousness has over human affairs, the more truly human, and hence super-human, man will become. Now the common ownership of the means of production implies the consciously planned control of production. No longer is production to be governed by the self-acting mechanism of profitability; it is to be carried on for communal use. No longer, at a given conjuncture of the world-market, will so many dozen factories automatically go out of action in some far corner of the earth, throwing some thousands of workers into immediate poverty, and diverting the energies of the owners into other channels. No longer will thermodynamic efficiency and geographic common sense alike be turned upside down at the irrational dictates of a profit-making system. By the deliberate decisions of a central planning body the production and distribution of goods will be consciously organised. Spencer might well have welcomed such a body as the real analogue of the higher nervous centres for which he had to seek in contemporary society in vain. Its rationalisation of the irrational in terms of practice will be the precise counterpart of the rationalisation of the irrational in terms of theory which the scientific method carries out in research every day.

But this vast extension of conscious control could not take place without the willing conscious co-operation of the constituent effector persons. In Vaughan's discussion of Rousseau there is a fine passage on the conception of the State.[2] "It is of the essence of Rousseau's theory," writes Vaughan, "that the State is no power which imposes itself from without; that, on the contrary, it is more truly part of the individual than the individual himself. The change he wrought in the conception of the individual involves a corresponding change in the conception of the State. The bureaucratic machinery, which had slowly fastened itself on Europe, is thrown to the winds. Its

[1] The processes of "rationalisation" in industry offer another parallel. When the American Railway Association reduces the number of types of axles from 59 to 5 numerical complexity decreases but organisation increases. *Mere* complexity has, of course, nothing to do with organisation—it may mean confusion only—and purposed reduction of complexity at a lower level is in the interest of activities at higher levels. Industrial standardisation is an example which would have appealed greatly to Herbert Spencer, the railway engineer.

[2] *The Political Writings of J. J. Rousseau*, edited with introduction and notes by C. E. Vaughan (Cambridge, 1915), p. 112.

place is taken by the idea of a free community, each member of which has as large a share in determining the 'general will' as his fellows; in which, so far as human frailty allows, the general will takes up into itself the will of all. . . . The only State he recognises as legitimate is the State of which the sovereign is the People." But it is evident that none of this service which is perfect freedom can be secured without unlimited universal education and the abolition of classes. The matter has been put more shortly: "Every cook," said Lenin, "must learn to rule the State."

Every transition from the unconscious to the conscious implies a step from bondage to freedom, from lower to higher level of organisation. All early agriculture and storage of food-products necessitated more conscious control than before. Increases in the efficiency of mechanisms of transport from the horse to the aeroplane widened men's conscious horizons. In the realm of the individual, modern psychology provides brilliant examples of the liberating effects of a passage from the unconscious to the conscious, e.g. in the cure of the obsessional neuroses. Up to the present all commercial transactions have been the instruments of a peculiarly subtle form of bondage which was called by Marx the "fetishism of commodities."[1] Relations such as those of exchange in the open market between commodities, appear at first sight to be relations between things, but are on the contrary relations between persons, the persons who produced them and the persons who will consume them. To forget this is to be forced to assent to the various "iron laws" of political economy, which have in reality nothing inescapable about them, once the personal relationship is grasped.[2] In one of the most inspired passages he ever

---

[1] "The commodity form, and the value relation between the labour products which finds expression in the commodity form, have nothing whatever to do with the physical properties of the commodities or with the material relations that arise out of these physical properties. We are concerned only with a definite social relation between human beings, which in their eyes, has here assumed the semblance of a relation between things. To find an analogy, we must enter the nebulous world of religion. In that world, products of the human mind become independent shapes, endowed with lives of their own, and able to enter into relations with men and women. The products of the human hand do the same thing in the world of commodities. I speak of this as the *fetishistic character* which attaches to the products of labour, as soon as they are produced in the form of commodities. It is inseparable from commodity production."—K. Marx, *Capital*, tr. E. and C. Paul (London, 1928), p. 45.

[2] The realisation of the entry of personal relationships at this point gives us the answer to the perplexity of Frederic Harrison the positivist in his Herbert Spencer lecture of 1905. He said that, while according to "evolutionary philosophy" the unceasing redistribution of matter and motion seemed to be the fundamental law, he himself

wrote,[1] Engels said that "the seizure of the means of production by society puts an end to commodity production and hence to the domination of the product over the producer. Anarchy in social production is replaced by conscious organisation on a planned basis. The struggle for individual existence comes to an end. And at this point, in a certain sense, man finally cuts himself off from the animal world, leaves the conditions of animal existence behind him, and enters conditions which are really human. . . . It is humanity's leap from the realm of necessity into the realm of freedom."

Lastly, it must be emphasised that our present civilisation is manifestly not a state of stable equilibrium. The enormous advances in scientific knowledge and practical technique, due themselves in a large degree to the middle-class economic system of which Spencer was the representative, have made that system an anachronism. Nothing short of the absolute abolition of private ownership of resources and machines, the abolition of national sovereignties, and the government of the world by a power proceeding from the class which must abolish classes, will suit the technical situation of the twentieth century.

After all, within single human lifetimes it has sometimes been possible to discern the advance of human society to ever more complex unities and higher levels of organisation. Towards the end of his noble autobiography,[2] that great American, Henry Adams, noted the two outstanding instances of the victory of the larger unit in his own experience, first, the triumph of the North in the American Civil War, secondly, the failure of the "trust-busting" period of legislation. "All one's life one had struggled for unity, and unity had always won." It may be illuminating to regard the present World War (1942) as a movement of *secessionism* just as surely as the Civil War which Adams lived through—not indeed from a World State already in being, but from the idea of the World State for which humanity is obviously ready. All dominating racialisms, at this stage of world

could be content with nothing which did not include mention of "progress," "order," "living for others," justice, love, etc. But these things are not something superadded to nature, they arise within it and grow out of it. Comradeship is precisely one of the essential conditions for high social organisation, and like all the other highest human qualities a natural product. After Bacon's time, said Marx (in the *Holy Family*) materialism became misanthropic and ascetic. It was rescued by the French and by Marx himself.

[1] F. Engels, *Anti-Dühring* (London, n.d.), p. 318.
[2] *The Education of Henry Adams* (first published by the Massachusetts Historical Society, 1907, also London, 1928), pp. 398, 402, 500.

history, are secessions from this wider idea. And secessionist minorities are bound to fail because, as Seversky[1] has put it, that side "with the greatest economic strength, industrial capacity, and engineering ingenuity will have the advantage, as always throughout history." We may agree with the words of a Russian philosopher; "However strong the forces of armed reaction, in the end progressive mankind has invariably found the strength to win the victory, and to preserve and develop the achievements of the human mind."[2]

Evolution and Inevitability.

Now if it has been shown that the organisation of human society is only as yet at the beginning of its triumphs, and that these triumphs are *inevitable*,[3] since they lie along the road traced out hitherto by the entire evolutionary process, is there not some danger lest the effect of such a belief should be to withdraw our own activity from the daily struggle for a better, because better organised, world? If collectivism[4] is inevitable, why not just sit and wait for it? There is

[1] A. Seversky, *Victory through Air Power* (London, 1940), p. 140.

[2] G. Alexandrov, Soviet War News, 2nd June, 1942. Cf. also "It will be convincingly demonstrated that those who invoked force in violation of their obligations to a world order were destroyed by the inherent capacity of the world order to invoke a greater force in its own defence" (2nd Report of Carnegie Organisation of Peace Commission, 1942, p. 163).

[3] "The downfall of the bourgeoisie and the victory of the proletariat are equally *inevitable.*"—Marx and Engels, *Manifesto*, 1848.

"Once you unbridle the forces" (of world-war) "which you will be powerless to cope with, then, however matters go, you will be ruined at the end of the tragedy, and the victory of the proletariat will either have already been won, or will in any case have become inevitable": Engels, preface to Borkheim's *Erinnerung*, 1887.

With these passages it is interesting to compare a christian formulation: "We believe that there is a purpose running throughout the whole universe, a purpose and a plan; and that if there be a purpose there must be a Purposer, whom we call God: that in spite of the at present inexplicable mystery of pain and cruelty, He is expressing Himself through the everyday virtues of the common people, through the heroic self-sacrifice and service of the saints, and through Jesus, the crown of humankind, God's word and energy who gives meaning and purpose to the age-long process. We believe that ... he opened the gate of heaven to all believers. This gate is not only a gate into a realm beyond death, but into a realm which descends and is incarnated in a fair, joyful, and equal commonwealth on the arena of this earth."—Conrad Noel, Church Militant, 1937.

[4] There exists a misapprehension in the minds of some, that the various forms of fascist government embody collectivism. This, however, has been completely exploded by many students, see e.g. G. Salvemini, *Under the Axe of Fascism* (London, 1936), and R. Pascal, *The Nazi Dictatorship* (London, 1934), or F. L. Schuman, *Hitler and the Nazi Dictatorship* (London, 1936), for Italy and Germany respectively. It is clear that fascism is a screen for the maintenance and stabilisation of existing class-stratification. The barbaric and militaristic tenets of fascism would be menacing indeed if we did not reflect that a relapse into barbarism seems to accompany each great transformation of economic

here a moral and psychological problem of considerable interest, to which a whole book has been devoted by Brameld.[1] How do real alternatives arise in the world-process, if the result is inevitable? To this question the answer of a representative communist spokesman would be of interest. We have it in a recent paper of R. P. Dutt, arising out of a controversy with the Dutch writer de Leeuw.[2] "It is the very heart," he says, "of the revolutionary marxist understanding of inevitability that it has nothing in common with the mechanical fatalism of which our opponents incorrectly accuse us. This inevitability is realised in practice through living human wills under given social conditions, consciously reacting to those conditions, and consciously choosing their line between alternative possibilities seen by them within the given conditions. 'Man makes his own history, but not out of the whole cloth.' We are able scientifically to predict the inevitable outcome, because we are able to analyse the social conditions governing the consciousness, and the line of development, of those social conditions. We are able to analyse the growth of contradictions, and the consequent accumulation of forces generating ever greater revolutionary consciousness and will in the exploited majority, till they become strong enough to overcome all obstacles, and conquer. We are able to lay down with scientific precision that every failure, every choice of an incorrect path, can only be temporary, because the outcome can in no way solve the contradictions generating the revolutionary consciousness and will. These contradictions then only lead to renewed and intensified struggle, up to final victory. The process is inevitable. But the human consciousness of the participants in this inevitable process is not the consciousness of automatic cogs in a predetermined mechanism. It is the consciousness of living, active, human beings, revolting against intolerable evils, deliberately with thought and passion choosing a new alternative, doing and daring all to achieve a new world, and ready to give their lives in the fight because of their intense desire to help by such actions to make possible the achievement of the goal. This fighting revolutionary consciousness is by no means a bowing to an inevitable outcome, but is most actively a seeking to tip the balance and make certain by

structure. When feudalism was giving place to middle-class capitalism there were the "wars of religion," and the witchcraft mania, even before the English civil war and the French revolution. The middle class will not consent to merge itself in the comradeship of mankind without similar catastrophes.

[1] T. B. H. Brameld, *Philosophic Approach to Communism* (Chicago, 1933).
[2] R. P. Dutt, Communist International, 1935, **12**, 604.

action the victory of one alternative and the defeat of another alternative. Every revolutionary worth his salt acts in every stage of the fight as if the whole future of the revolution depended on his action. And in presenting the issues of the present day to the masses, we present them not as placid inevitabilities to contemplate like the movement of the stars, but as a gigantic issue with the whole future of humanity at stake, calling for the utmost determination, courage, sacrifice, and will to conquer. This is the essence of the revolutionary marxist understanding of inevitability."

From this passage two principal thoughts emerge. In the first place, the marxist writer speaks of inevitability only because he is confident that he understands the nature of human beings and their reactions to external conditions. Their knowledge of good and evil, pleasure and pain, will have its ultimate effect and that effect is inevitable.

We can take our stand upon the simple, natural, healthy, human desires of the mass of mankind, for love, for children, for socially useful work, for fundamental decency and dignity. This is the meaning of the ancient Confucian advice to rulers, in the *Ta-Shioh* (Great Learning)—*Min chih so hao, hao chih; min chih so ô, ô chih*; "Love what the people love, and hate what the people hate."

If we explore more closely the mechanism of this inevitability we see that it is connected with the contradictions which arise in each successive stage of human history. Thus modern nationalist states must arm their workers in their struggles with foreign imperialisms, yet at the same time this is to arm their destroyers. They must engage in colonial development, but this gives rise to native movements of liberation. In the last resort fascist theory is brought in to save the decaying structure, and this essays to substitute for Reason a fantastic irrational mythology, but on the other hand modern capitalism cannot get on without effective control over nature, and this necessitates scientific rationality.

Secondly, inevitability once admitted, the time-scale remains only too obscure. It is true that we might envisage a long period of stagnation as the outcome of our present civilisation. China is sometimes thought (without much justice) to present a century-long spectacle of such stagnation. But whereas this might be compatible with an agricultural, bureaucratic, isolated community lacking good communications and so able to sterilise revolutionary movements within itself, it is much more difficult to imagine such a state of affairs existing in a civilisation based on scientific technology. Let us grant, however, that some kind of scientifically stabilised stagnant class-stratified totalitarian social organism, might succeed our own age. Hence the great significance of the word *"temporary"* used in the passage from the writer quoted above. Failures and set-backs and blind alleys there may be in plenty, but though the ultimate victory is not in doubt, it must be remembered what each failure may mean. It may mean the enslavement of whole peoples for many generations, the destruction of culture and learning over a wide part of the world, the stagnation of social progress in such regions, the martyrdom of many thousands of our best and noblest friends. In the ancient phrase: "the saints under the altar cry, O Lord, how long, how long?" To speak of the inevitability of our higher integrative level is to say nothing of when it will come.[1]

Conclusion.

It would be a pity, however, to conclude this lecture upon a note of sadness. Let us return to the year 1838, when Herbert Spencer was a young man of seventeen. The youth of anyone so exsuccous as Spencer was in his old age has always a peculiar charm. This was the year in which Marx was toiling in Berlin at his doctoral dissertation on Democritus and Epicurus, Engels was quietly acquiring a business training at Bremen, and Darwin, just back from the voyage of the *Beagle*, was starting his first Notebook on the Transformations of Species.[2] Spencer, as befitted his later outlook, was in the midst of British industry. Under Mr. Robert Stephenson, the chief engineer of the London and Birmingham Railway, young Mr. Spencer made measurements of embankments and cuttings, drafted out plans, and sketched minor inventions in his spare time. To every man who much

---

[1] Cf. the interesting analysis of causality and determinism by H. Levy, Proc. Aristot. Soc., 1937, **37**, p. 89. "Such a form of analysis," he concludes, "will tell us how the causal process operates, and, in terms of the qualities of subsidiary group-isolates, when a dialectical change will occur. It cannot express the prediction in terms of time." So also the conclusion of an acute student of the history of science—"Great men are not absolutely essential to the progress of science, but they increase its speed." (J. G. Crowther, *The Social Relations of Science*, London, 1941, p. 453.)

That moving play of Robert Ardrey's, *Thunder Rock* (London, 1940) gave brilliant expression to this (see esp. around p. 117).

[2] Comte, with his conviction that philosophy must acquire a social relevance, his appreciation of social evolution before biological evolution was substantiated and accepted, and his realisation that the classification of the sciences concealed a real problem, deserves a lecture to himself. At this time he was publishing his *Positive Philosophy*, the first volume of which appeared in 1839 and the last in 1842.

affects his fellows, there comes at one time or another in his life a symbolic event, and some sixty-five years later Spencer described it in language which insists upon quotation.[1]

"Harris and I were sent down one day early in August to make a survey of Wolverton station, and we completed it before evening set in. Wolverton, being then the temporary terminus, between which and Rugby the traffic was carried on by coaches, was the place whence the trains to London started. The last of them was the mail, leaving somewhere about 8. If I remember rightly there were at that time only five trains in the day and none at night. A difficulty arose. The mail did not stop between Watford and London, but I wished to stop at the intermediate station, Harrow, that being the nearest point to Wembley. It turned out that there was at Wolverton no vehicle having a brake to it—nothing available but a coach-truck. Being without alternative, I directed the station-master to attach this to the train. After travelling with my companion in the usual way till we reached Watford, I bade him goodnight, and got into the coach-truck. Away the train went into the gloom of the evening, and for some six or seven miles I travelled unconcernedly, knowing the objects along the line well, and continually identifying my whereabouts. Presently we reached a bridge about a mile and a half to the north of Harrow station. Being quite aware that the line at this point, and for a long distance in advance, falls towards London at the rate of 1 in 330, I expected that the coach-truck, having no brake, would take a long time to stop. A mile and a half would, it seemed, be sufficient allowance, and on coming to the said bridge, I uncoupled the truck and sat down. In a few seconds I got up again to see whether all the couplings were unhooked, for, to my surprise, the truck seemed to be going on with the train. There was no coupling left unhooked, however, and it became clear that I had allowed an insufficient distance for the gradual arrest. Though the incline is quite invisible to the eye, being less than an inch in nine yards, yet its effect was very decided; and the axles being, no doubt, well-greased, the truck maintained its velocity. Far from having stopped when Harrow was reached, I was less than a dozen yards behind the train! My dismay as we rushed through the

[1] A, I. 134.

station at some 30 miles an hour may be well imagined.... After passing Harrow station the line enters upon a curve, and a loss of velocity necessarily followed. The train now began rapidly to increase its distance, and shortly disappeared into the gloom. Still, though my speed had diminished, I rushed on at a great pace. Presently, seeing at a little distance in front the light of a lantern, held, I concluded, by a foreman of the plate-layers, who was going back to the station after having seen the last train pass, I shouted to him; thinking that if he would run at the top of his speed he might perhaps catch hold of the waggon and gradually arrest it. He, however, stood staring; too much astonished, even if he understood me, and as I learned next day, when he reached Harrow, reported he had met a man in a newly-invented carriage which had run away with him!... After being carried some two miles beyond Harrow, I began rather to rejoice that the truck was going so far, for I remembered that at no great distance in advance was the Brent siding, into which the truck might easily be pushed instead of back to Harrow. I looked with satisfaction to this prospect, entertaining no doubt that the waggon would come to rest in time. By and by, however, it became clear that the truck would not only reach this siding but pass it; and then came not a little alarm, for a mile or so further along was the level-crossing at Willesden, where I should probably be thrown out and killed.... However, on reaching Brent bridge, the truck began to slacken speed, and finally came to a stand in the middle of the embankment crossing the Brent valley."

How Spencer had to seek help to clear the line and finally got home in the early hours of the morning, we need not here relate. But of all the symbolic occurrences which have happened to great men, this is surely one of the most remarkable. Spencer wanted to stop at the intermediate station in evolutionary sociology, but in the progress of organisation to ever higher levels, there is no such opportunity. The class of which he was the intellectual representative wanted to stop at the intermediate station of domestic capitalism, but the inner logic of the process demanded that expansion should go on and the local mill-owner should give place to the trustified imperialist. Moreover, the inevitable industrialisation of the working-class led to demands of diametrically opposite nature, so that Spencer was driven into the position of protesting vainly against both the "degeneration" into militarism, and the socialist movement which fought against "free competition." His wide, and substantially correct, survey of evolution led up only to the anti-climax of middle-class liberal economic individualism, past which, in spite of himself, he was carried on protesting.

But let us celebrate his noble range of vision nevertheless. The onward progress of integration and organisation cannot be arrested. As I write, there rages in one of the most beautiful of European countries a tragic and terrible struggle between the People and their Adversary. The sound of its gunfire penetrates any College court, no matter how peaceful it may seem. Some faith may be needed to assert with boldness that, even if Spanish democracy be overwhelmed, even if the great democracy of the Soviet Union itself were to be overwhelmed, no matter what shattering blows the cause of consciousness may receive, the end is sure. The higher stages of integration and organisation towards which we look have all the authority of evolution behind them. It is no other than Herbert Spencer himself who contributes to this our faith, if faith it be. The devil, as Hippolytus said long ago, may resist the cosmic process. But the last victory will not be his.

Cambridge,
April 16, 1937

INDEX

Abelard 19
Adams, Henry—
    on the Virgin of Chartres 49
    on the victories of the larger unities 265
Adler, M. J. 15, 18, 192
Aeschylus 27
Afinogenov 66
aggregation and disaggregation 39, 191
agitators 80
Agreement of the People, The 80, 101
Albertus magnus 151
Albigensians 239
d'Alembert 123
Alexander, Samuel 112, 185, 234
    on beings in social relation 125
    on embryonic determination 197
    on time 236
    on mind as a property of high organisational level 259
Alexandrian science 144, 146
Alexandrov, G. 266
altruism, social, origins of 36
Amos 51
Anabaptists 239
Anaximander 27
Andrewes, Lancelot 75, 77, 90, 98
Angela of Foligno 238
Aquinas 31, 152
    and Marx 90
Archimedes 48
Ardrey, R. 269
aristocratic thought and popular thought, 27
Aristotelianism—
    account of causation 85 ff
    twilight of 103 ff, 110 ff
Aristotle—
    on the exercise of the spirit 22
    on succession of souls 32, 114
    on social classes 114
    and practical science 144
    audacity of 150
    on form and matter 153
    on the ladder of beings 187
art—
    and the numinous 60
    and religion 61

atomism—
    and capitalism 86 ff
attraction and repulsion 39, 191, 230
Auden 15, 24, 30, 31, 35, 59, 60, 65, 231
Augustine 51, 53, 138
authority and freedom 96

Bacon, Francis 105
Bagehot 114
Baker, J. R. 94, 109
Balfour, Arthur 236
Balfour, Frank 158
Ball, John 239
Barcroft, Sir Joseph 213
Bateson 250
Baxter, Rich 231
Bellers 88
Benda 46
Berdyaev 72, 240
Bernal 62, 95, 190, 228, 244
Bernard, Claude 213
Biggs, Noah 103
biochemistry 17, 174 ff, 205
biological engineering 71
biologism (McBride) 161
Blake 70
Blok 59
du Bois Reymond 8
Bolshevism, Christian theology the grandmother of 70
bondage to space-time 67
bondage to space-time 67
Bonnet, Ch. 154 ff
Borkenau, F. 169, 257
Boyle—
    on heaven 13
    on the aims of the Royal Society 84
    on atomism 87, 149
    and scientific technique 147
    on *a priori* and *a posteriori* arguments 155
Broad 196, 242
Browne, Sir T. 75
    on man's amphibious nature 8
    on heaven 14

Browne, Sir T. (*continued*)—
    on the mutations of the world 15
    on life and death 66
    on order 232
Bruno 13
Bukharin—
    on social man 23
    on organisational levels 160, 243
    on the history of science 54
Burnet 105
Butterfield 19
Byzantine Christianity 68 ff

Cambridge and the English civil war 99
Cannon, W. B. 115
capitalism, rise of 75
    glittering prizes of 43
    and atomism 86 ff, 186
    deficiencies of 88
    and imperialism 257
caste and fascism 71
Chase, S. 134
Chaucer 47
chiliasm 50
Chinese concepts—
    ch'êng ching, ataraxy 9
    li, right social behaviour 23
    vjêng, love of mankind 23
    i, social justice 23
    chwing dże, the scholar-hero 34
    tżn-rjan, Nature coming into being by itself 55
    teh, virtue 119
    chih, "stopping" at the highest moral good 119
Chrysostom 261
Church of Jerusalem—
    communism of 239
    distribution of 53
Cicero 113
Civil War—
    in 17th-Century England 19, 78 ff
    American 265
class-stratification of society 261
Clement of Alexandria 19, 55, 64
Coleridge 187
Collingwood 9
Comenius—
    on having a broad outlook 26
    on education 111
committees, divine character of 14

Commonwealth, the 75 ff
competition—
    intra-specific 251
    capitalist 88
complexity, its relation to organisation 263
Comte 269
Confucianism—
    and non-supernatural morality 10
    and social evolution 52
    and the essential goodness of man's nature 139
    its advice to rulers 268
Confucius 27, 56, 139
Contakion 66
contradictions 13
Cornford, John 34
Cosin, John 75
Crashaw 24, 75
creatureliness 65, 172
Cromwell 78 ff, 101, 257
Crowther, J. G. 107, 165, 217, 269
Cyprian 50

Dale, Sir Henry 94
Dalton, John 87, 199
Darlington, C. D. 94
Darwin 14, 93, 115, 163, 199, 269
Day Lewis 6, 24, 233
death, the problem of 66
Deborin 87
Dell, Wm. 99 ff
democracy, form of society called for by biology 164, 250
Descartes 151, 200, 202, 203
determination, embryonic 196
devil, the, definition of 247, 272
dialectics—
    in scientific thought 16
    in history 19, 104
    in evolving nature 189 ff
Diderot 123, 139
Digby, Sir Kenelm 85 ff, 151
Diggers 78 ff, 101 ff
Dimitrov 125
Dingle on *deus ex aequatione* 131
Dionysius the Areopagite 8
distortion of scientific ideas for reactionary political purposes 117
dividers and uniters 10, 27, 28
Dollo 222

INDEX

Donatists 239
Donnan—
    on integro-differential equations 213
    on magnification of disentropic processes 215
dragons 39
Driesch 179, 180, 196, 200, 216
Drummond 28 ff
    on the ascent of man 34
    on "war" and "industry" 37
Duns Scotus 123

east and west 136 ff
Eckhardt 13
economics—
    rise of mechanistic 85 ff
    domination of, over eugenics 166 ff, 262
    and theology 46
Eddington 112, 209 ff, 221
Ehrenburg 60
Eisenstein 60
Eliot, T. S. 16, 67, 172
embryology 17, 18, 107, 123, 143, 147, 148, 149, 157, 245
emergence 183
Empedocles 230
empiricism and rationalism 156
enclosures 76
Encyclopaedists, the 138 ff, 142
Engels 112, 185 ff, 255
    on contradictions 14, 19
    on fixed boundary-lines 20
    on levels of organisation 31
    on the philosophy of scientists 112
    on Darwinism 115, 163
    on organisation and mind 123
    on the dialectics of nature 190
    on the history of science 198
    on form and matter 206
    on evolution and entropy 214
    on the peasant wars 239
    on freedom and necessity 265
    his youth 269
engrossing 47
enmeshing the ideal in the real 119
entropy 207 ff
"envelopes" in space 184
Epicureanism—
    ataraxy 9, 49
    and mind 20, 124

Epicureanism (*continued*)—
    and the spontaneity of the world-process 55
    and practical science 144
episcopacy the outwork of property 79
Eros, the task of, in social progress 25
ethics—
    and politics 10
    scientific 55
    of a machine age 132 ff
eugenics, and economics 166 ff
evil, problem of 65
evolution 30, 31
    its methods and results 36
    the giant vista of 258
    social 260
        continuous with biological evolution 160, 235
        and thermodynamics 230
experience, forms of 8, 61
Ezekiel 119

Faraday 26
Farrington, B. 20, 113
fascism 266, 268
Fernelius 201
Ferrar, Nich. 75
Ferri, Enrico 254
fetishism of commodities 264
feudalism 48
Feuerbach—
    on the Trinity 14
    on social man 23
    on religion and art 61
Fisher, R. A. 162
folk art 128 ff
forestalling 47
form and matter 205
Franklin 93
freedom—
    and authority 96
    as the knowledge of necessity 97
Freind 155
French Revolution 20, 138, 241
furies 39

Galen 147
Gassendi 87, 151
Gautama 27
genius and feeblemindedness 162
George, Henry 253

Gibbs, Willard 169, 209 ff
Glanville 110
Glisson 206
Glover, E. C. (psychologist) 173
Glover, T. R. (historian) 237
Goethe 19, 152, 229
gospels, theology of the 66
Gotch, F. 110
Graubard, M. 167 ff
Gresham, Sir Tho. 87, 99
Grierson 60

Hacket, John 75
Haeckel 8
Haldane, J. B. S. 44 ff, 93, 162, 190
Haldane, J. S. 121 ff, 204, 227
Hall, Joseph 77
Hardy, G. H. (mathematician) 55
Hardy, W. B. (biologist and physicist) 179, 206
Harrison, F. (positivist) 264
Hartlib 88
Hartsoeker 155
Harvey 98 ff, 143, 148, 151, 153 ff, 157, 206
Headlam 58
Hegel 14, 187 ff, 190
Hellenistic—
    mysticism 51
    science 144
    world, stagnation of 237
Henderson, L. J.—
    on the fitness of the environment 258
    on Pareto's sociology 169
Herbert, Geo. 75
Herder 233
Hesiod on the golden age 52
Hessen, B. 107, 145
Heylyn, Peter 76
Hickes, Geo. 114
Hill, A. V. 44 ff, 92 ff
Hippolytus 247, 272
His, W. 158
Hobbes 123
Hogben 95, 167, 244
Holorenshaw, Henry 78
Hooke, Robt., and the Royal Society 84, 92
hookworm, its point of view 223, 236
Hopkins, Sir F. G. 179
Hsün-tze 139

Huxley, Aldous 71
Huxley, Julian 23, 57, 163, 251, 258
Huxley, T. H. 8, 26, 28, 35, 179, 181

I-Ching, the (Book of Changes) 52
ideals, unrealisable 119
Indian philosophy, probable origin of asceticism and otherworldliness 238
individual, importance of 33
individualist fallacy 23, 203
"inevitability" of social evolution 266 ff
Inge 25, 164, 218, 237 ff, 240
Inquisition, the 70
Invisible College, the 98 ff
Irenaeus 50, 52, 64, 125
Isaiah 237, 238

Jalal'ud-Din Rumi 32
Jeans, Sir J. 112
Jennings, H. S. 162, 179, 258, 261
Jeremiah 238
Jesuits 139
Jewel, John 77
John the theologian 50
Justin 50

Kautsky 240
Ken, Tho. 75
Kierkegaard 65 ff
King, Henry 75
Kingdom of God, what 43, 50 ff, 89, 239
Klages 136
Kuan-tze 97
Kung-fu-tze see Confucius
Kuo, Z. Y. 123
Kyrle, R. M. 72, 173

Laud, Wm. 75 ff
Lawrence, D. H. 62
Lenin 34
    on dialectics 19, 244
    on the Machians 21
    on religion 57
    on revolutions 69
    on organisation and mind 123
    on Marx 255
    on cooks 264
    and Lloyd-Morgan 112

INDEX

Leonardo da Vinci 147
Lessing 22
Levellers 19, 75 ff, 78 ff, 90, 125, 240
levels of organisation 30, 31, 122, 211, 233 ff
    and mind 122 ff, 259
    integrative 233 ff
Levy, H.—
    on evolution and entropy 214
    on dialectical change and time 269
Lewis, G. N. 215
Lewis, John, on the sacred and the secular 118 ff
Liebig 38
Lilburne 34, 64, 80
Lillie, Ralph S. 215, 229
limiting factors in scientific discovery 141 ff
liquid crystals 235
Lloyd-Morgan 112, 185, 225
Lockyer, Robt. 82
Loeb, Jacques 179
Lollards 239
Longus 39
Lotka 222
Lotze 20
Lucretius 20, 71
    on Venus as the goddess of union and aggregation 40
    on science freeing the mind from fear 49
    on the spontaneity of the world-process 55
    on the "swerve" of the atoms 124
    on the dialectics of the old and the new 191
    on mind as a property of high organisational level 259
Lunacharsky 69
Lunar Society, the 106

McTaggart 126, 132
Mach 21
machinery, contacts between men and 127, 134
Mallock, W. H. 179
Malthus—
    on population 38
    on improper arts 164
Malynes, Gervase 86

manichaeism 131 ff
marriage 173
Marson 64
Marvin, W. T. 180
Marx 27, 31, 53, 112, 132, 170, 185 ff, 203, 255
    on contradictions 14, 19
    on Liebig and Malthus 38
    on changing the world 65
    on religion 118
    on Spencer 255
    on the fetishism of commodities 264
    his youth 269
    and Darwin 14
materialism—
    mechanical 20, 122 ff, 188
    "misanthropic" or "ascetic" 188
    in Christianity 52, 122, 126 ff
    dialectical 14, 122 ff, 186 ff
    and God 55
Maublanc 14
Maudsley—
    on social man 23, 125
    on mind as a property of high organisational level 259
Mayakovsky 59
Mead, Margaret 175
mechanics, statistical 208 ff
mechanisation of democracy 164
mechanism (biological) see vitalism
Mencius 139 ff
Mendeleev's table 225
Mêng-tze see Mencius
Merejkovsky 67
mesoforms 235
metabolism and thermodynamics 220 ff
Meyerhof 225, 228
Miall, L. C. 98
Mill 65
millenniarism 51
    and social evolution 240
Milne, E. A. 219
Milton 103, 110 ff
mind and matter 201
mixed-up-ness and separatedness 209 ff
moral theology, of to-day 54
More, Henry 75
More, Sir Thomas 34
morphology 205, 245
"morpholysis" 211
Muller, H. J. 113, 166 ff
Murry 57

natural and supernatural 29, 200
natural selection 36
    and "laissez-faire" 116, 163
naturalism, scientific 20
nature—
    and grace 31
    human, not unchangeable 175
nazism 62, 266
    and biology 163
Needham, Jasper 145
Needham, John Turberville 157
Needham, Joseph (18th century) 143
Needham, Marchamont 103
Newton 70, 99, 107, 145
Nicholas of Cusa 13
Noel, Conrad 14, 237 ff, 266
numinous, the—
    in religion and art 60
    defined 64, 125
    and socialist emotion 66 ff
    and the social order 120

obstetrics 143
opium—
    religious 65, 113
    scientific 67 ff
order—
    and arrangement 208 ff
    and organisation 220 ff
organicism 242 ff
organisation—
    levels in evolution 30, 31, 32, 122, 185, 192 ff, 211 ff, 233 ff, 258 ff
    and energy 33, 198, 206
    and mind 122 ff, 201, 259
    and entropy 207 ff
    definition attempted 211, 258
    the two concepts of 213 ff
Origen 51
original sin 113, 138
Orphism 27, 39
Orr, Sir J. B. 176
Orwell 25
Oparin 191
Otto 64, 124 ff, 132
Overton, Rich. 80 ff, 113
Owen and the Chartists 88

Papias 52 ff
Pareto 168 ff
Parker, Samuel 85

Pascal, R. 66
Pasteur 26, 34, 199
patent medicines 116
pattern in nature 212 ff, 224 ff
Pavlov—
    on the subjective and the objective 123
    on manual and mental work 156
    on human nature 175
    and mind 202
Pelagius 138
Pepys 93
pessimism, post-Victorian 236
physicism (Pareto and Bogdanov) 168 ff
Pirenne 257
Planck 23, 223
Plato 27, 114, 164
Pledge, H. T. 107
Plekhanov 55, 61
Plockboy 88
Plotinus 237
poetry—
    and science 25
    of the liturgy 58
Polanyi, Karl (theologian), on regressive movements in thought 136
Polanyi, Michael (physical chemist)—
    on capitalism and atomism 87
    on the holiness of pure science 94 ff, 120
Ponnet, John 98
population control 164, 171
Potter, Beatrice (Mrs. Sidney Webb) 254
Prenant 189, 244
Presbyterians and Independents 78 ff
Priestley 93
Pritchett 24
protestantism, capitalism and science 98
psychology and psycho-analysis 171 ff

Quakers 81

racialism 63, 116
Rainborough 34, 64, 80
Rashdall 25, 219
Redi 142
"reducibility" of biological fact 204
regrating 47
religion—
    as the sense of the holy 23

INDEX

religion (*continued*)—
    in the 17th century 42 ff
    and organised religion 57 ff
    of the future 59
    and art 60
    influences antagonistic to 63
    as the opium of the people 65
    evolution of 126 ff
    as the mirror of exploitation 127
Roberts, Morley 115
Romanes 35
Rousseau 139, 142, 263
Roux, W. 157, 181
Rowse 65
Royal Society, the 83 ff, 92 ff
    political background of, in the 17th century 97 ff, 101
    industrial associations of 104 ff
Russell 71 ff, 223

sacred, the, and the secular 43, 120
Scarborough, Charles 98
Schmalhausen, S. D. 172 ff
Schrödinger 220
science—
    influences antagonistic to 62
    and fascism 73
    and the rise of capitalism 84, 90
    pure and applied 84, 92 ff, 106, 107, 108, 109
    ethical neutrality of 92
    and the idea of the holy 94 ff
    theoretical, in a socialist society 108
    in Russia 108
    and planning 109
    and philosophy 110
    and empirical technique 143
    and theology 145
    and terminology 148
"scientific" socialism 188
scientific societies in the renaissance 145
scientific worker, position of to-day 43 ff
secessionism 265
Sechenov 26, 123
Sellars, Roy W. 112, 185
Seversky, A. 266
Sex 173 ff
Shaw 12
Sherrington, Sir Charles 184, 202, 206
Shirokov 189

similarity of the works of organisation at different levels 212
Singer, Ch. 107
Smuts 112, 185
social emotions, as bonds compared with interatomic forces 23, 35, 39, 56
society—
    as an organism 113, 114 ff, 249
    aim of 117
sociology—
    a level of its own 33, 160 ff, 176
    boundaries fixed by other sciences 176
Spaun, Othmar 115, 135, 240
Spemann 148
Spencer 28, 29, 31, 114, 233 ff
    on embryonic development 245, 248
    on evolution 246
    on the world-process 247
    on organisms and environment 247
    on sociology 248 ff
    on "predatory" and "industrial" societies 251
    on socialism 252 ff
    on social success 262
    his youth 269
    his adventure on the L. & B. Rly. 270
Spengler 72, 249
Spinoza 55
Sprat, Tho., and the Royal Society 84, 104, 145
Stalin—
    on the old and the new 191
    on organic mechanism 192
    on Lenin 204
Stamp, Josiah 135
Stebbing 58
sterilisation, eugenic 162
Strasser, Otto 240
"succession" in time 184
superman, doctrines of 73
surplus value 256
survival of the fittest 115, 262

Ta-Shioh, the (The Great Learning) 119, 138 ff, 268
Taborites 239
Talmud, the 146

Taoism—
    and peace of mind 9
    and the spontaneity of the world-process 55
Taylor, Jeremy 75, 98
Tawney 77
technocracy 117
Temple, Wm. 126
Tertullian 50, 64
theology, Christian, the grandmother of Bolshevism 70
thermodynamics 207 ff
Thomson, Geo. 27, 217
time, importance of 236 ff
Timiriazev 26, 67
treason of the scholars 46 ff
Tyrrell 60

Unwin, J. D. 173
Usury 47, 76 ff

Vaughan, C. E. (editor of Rousseau) 263
Vaughan, Henry 75
Venus, goddess of union and aggregation 40
vitalism and mechanism 18, 32, 160, 178 ff, 184, 241 ff
Voltaire 139 ff

Waddington—
    on science and ethics 23
    on Huxley's attitude to Nature 36
    on Polanyi's conception of science 95
    on authority and freedom 96
Waldensians 239

Walwyn, Wm. 80 ff
wars of religion 267
Watson, David 226
Whichcote 99 ff
Whitehead 112, 178 ff, 225
    on organisms and evolution 192 ff
    on organic mechanism 195 ff
    on time 198
    on the history of science 199
    his dialectical character 200
    on nature and mind 201
    and the social situation 203
    and metaphysics 194
William of Champeaux 19
Willis, Tho. 104
Winstanley, Gerrard 81 ff, 101 ff, 113
witchcraft 267
Wolff, Caspar Friedrich 142
Wolff, Christian 140
Woodger 182, 192, 242 ff
Wootton 74
workers, mystification of, in factories under capitalism 117
world State, arising out of the evolutionary process 41
Wren, Mathew 75
Wyclif 239

Yeats—
    on the thoughts of the wise in the language of the common people 24
    on modern English poetry 25

Zavadovsky 189, 244
Zechariah 119
Zoology 143
Zuckerman 260